BLOOMSBURY

Concise

Butt
& Moth
Guide

There are 47 individual Wildlife Trusts covering the whole of the UK and the Isle of Man and Alderney. Together The Wildlife Trusts are the largest UK voluntary organization dedicated to protecting wildlife and wild places everywhere – at land and sea. They are supported by 791,000 members, 150,000 of whom belong to their junior branch, Wildlife Watch. Every year The Wildlife Trusts work with thousands of schools, and their nature reserves and visitor centres receive millions of visitors.

The Wildlife Trusts work in partnership with hundreds of landowners and businesses in the UK. Building on their existing network of 2,250 nature reserves, The Wildlife Trusts' recovery plan for the UK's wildlife and fragmented habitats, known as A Living Landscape, is being achieved through restoring, recreating and reconnecting large areas of wildlife habitat.

The Wildlife Trusts also have a vision for the UK's seas and sea life – Living Seas, in which wildlife thrives from the depths of the oceans to the coastal shallows. In Living Seas, wildlife and habitats are recovering, the natural environment is adapting well to a changing climate, and people are inspired by marine wildlife and value the sea for the many ways in which it supports our quality of life. As well as protecting wildlife, these projects help to safeguard the ecosystems we depend on for services like clean air and water.

All 47 Wildlife Trusts are members of the Royal Society of Wildlife Trusts (Registered charity number 207238). To find your local Wildlife Trust visit wildlifetrusts.org

BLOOMSBURY

Concise
Butterfly
& Moth
Guide

BLOOMSBURY
LONDON · NEW DELHI · NEW YORK · SYDNEY

First published in 2010 by New Holland Publishers (UK) Ltd
This edition published in 2014 by Bloomsbury Publishing Plc

Bloomsbury Publishing Plc, 50 Bedford Square, London WC1B 3DP

www.bloomsbury.com

Bloomsbury Publishing, London, New Delhi, New York and Sydney

A CIP catalogue record for this book is available from the British Library
Library of Congress Cataloging-in-Publication Data has been applied for

Design by Alan Marshall

ISBN (print) 978-1-4729-0996-1

Printed in China by Leo Paper Group.

This book is produced using paper that is made from wood grown in managed
sustainable forests. It is natural, renewable and recyclable. The logging and
manufacturing processes conform to the environmental regulation of the country
of origin.

10 9 8 7 6 5 4 3 2 1

Contents

Introduction

Butterflies are the most obvious of insects and probably the most popular. They are seen during the day, mostly in the warmer months from spring through summer to autumn. Many moths are nocturnal, and are often seen when they enter the house or come to a lighted window. There are, however, some day-flying species that are so colourful people think they are butterflies.

Main Characteristics of Butterflies & Moths

Butterflies and moths belong to a huge order of insects, known as the Lepidoptera, which has over 165,000 species worldwide, with 2,300 in the British Isles. Moths are much more numerous than butterflies, and of the British species of Lepidoptera only about 70 are butterflies. Butterflies and moths are distinguished from other insects by being densely covered with tiny powdery scales ('Lepidoptera' is derived from the Greek 'scaly wings'). Sizes vary from species that are so tiny that a magnifying glass is needed to see them, to some tropical species that are the size of small birds.

Body Structure

Butterflies and moths have bodies that are in three parts:

• The head carries the eyes, antennae and mouthparts.
• The thorax has three segments. The first segment carries the front legs, the second the forewings and the middle pair of legs, and the third the hindwings and the third pair of legs.
• The third segment of the body is the abdomen, which has no legs, but does have sexual and digestive functions.

The Senses

Vision is through two large compound eyes and a pair of *ocelli*, or simple eyes. The *antennae* are found between the eyes. They are complex sense organs that can pick up chemical and tactile messages. Butterflies have antennae that are more uniform than those of moths, which are very varied.

Feeding

The vast majority of butterflies and moths feed on nectar and other liquids. They lack jaws, which have evolved into tongues. The tongue or **proboscis** is long and slender enough to probe flowers. When not in use it is coiled up and cannot be seen. The larvae or caterpillars feed on plants, though some members of the Lycaenidae also feed on ant grubs.

The Role of Colour in the Wings

The wings of butterflies and moths may be very strikingly coloured and patterned, or quite dull when they serve a protective role. The bright metallic colours, particularly purples and blues, are due to the structure of the **scales**, and if they are damaged the colours fade.

The bright colours must play a role in communication between individuals. Males also have specialized **scent glands** in the wings. The pheromones that they produce help to attract females over a distance, and also help to differentiate between the sexes of species in which males and females look alike.

The wings of some species may have roughly circular markings that look like eyes. It is thought that these distract predatory birds from the body of the insect. It may survive if a bird pecks a chunk from the wing, whereas a peck to the body will almost inevitably be fatal.

A protective mechanism employed by moths is **flash colouring**, where the forewings are well camouflaged, but the underwing is brightly coloured. When the moth is disturbed the bright yellow of the large yellow underwing will confuse a bird, because the moment the moth comes to rest the yellow disappears.

The upperside of a butterfly's wings may be brightly coloured, but the underside, which is the area visible when the insect is at rest, may be extremely well camouflaged, so that the resting animal looks like a dead leaf.

Temperature Control

Wings help to warm butterflies to the temperature that they need to fly. Their bodies need to reach about 30ºC before they can take off. The dark-coloured areas of the wings absorb the most heat, even on cool days when there is some sunshine. To achieve this temperature butterflies sunbathe with their wings open. Nocturnal moths reach the required body temperature by shivering their wings.

The Life Cycle

The life cycle of all insects includes a number of stages. These vary between species in terms of the timing, but each cycle is timed to provide the maximum opportunity for the larvae to feed. All butterflies and moths pass through four very different stages, during which they metamorphose completely.

• The first stage is an *egg*. The female lays an egg, in most cases on the plant on which the caterpillar will feed. The eggs are tiny and usually overlooked by the human eye.
• From the eggs hatch the *larvae* or *caterpillars*. These look nothing like the adult insect. They are worm-like with biting jaws, and in addition to the three pairs of legs on the thorax have stumpy legs on the abdomen. Caterpillars moult four or five times during their life. Before the final moult the caterpillar, which has become large and fat, seeks a place of safety, often burrowing beneath the soil, or protected within a silken cocoon that the larva spins, or hanging from its food plant. The caterpillars of some species – the hawkmoths for example – are large and spectacular.
• The final moult is the metamorphosis between caterpillar and *chrysalis* or *pupa*. The pupa has a harder outer skin than the larva and barely moves at all. However, inside, its tissues liquefy and it develops into the adult insect. The time for this to happen varies. In some species in warm weather it may be as short as a week, while in many it may take two to three weeks and in some several months, because

it is in this stage that the species spends the winter. The longevity of butterflies and moths varies between species and broods. Many species produce one brood a year, but others may produce more. The first butterfly to appear in early spring in Britain is the Brimstone, which hibernates as an adult. Males emerge first and patrol their territories. The greenish-white females appear a couple of weeks later.

Wingless Moths
The females of some moth species, such as the March Moth (*Alsophila aescularia*), do not have wings and are flightless. They can be seen crawling up tree trunks, on which they lay their eggs, at night in March–April; males can also be found on the wing at this time. March Moths are quite common in Britain, and their larvae feed on a range of deciduous trees, including oaks, hawthorns and fruit trees.

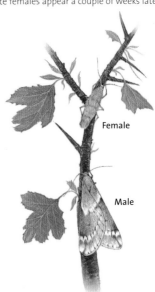

Female

Male

Eggs are laid in May and June, the larvae are seen in June and July, and the adults emerge in August. Sometimes among double-brooded species, some of the second brood will reach adulthood before winter and hibernate, while others will spend the winter as larvae or pupae.

Moths and butterflies are both seen seasonally. A common garden butterfly is the Peacock, especially if there is a bed of nettles on which the caterpillars might feed. The first Peacocks are seen as the weather warms in March and they emerge from hibernation to breed. After May they seem to disappear – this is because the adults have bred and the new generation is developing from eggs to larvae to pupae, emerging as butterflies from July to September. On sunny autumn days they can be seen feeding on the nectar or juices of rotting windfall fruits, before hibernating throughout the winter. You might come across one in a corner of a garden shed or even in your home (if you do, do not disturb it or put it outside in freezing weather).

BUTTERFLIES

Butterflies can be distinguished from moths mainly by the different shape of their antennae, which are usually thin and slender, and club shaped at the ends. At rest butterflies tend to hold their

wings vertically above the body, so that the underside is visible, while moths usually rest with their wings spread out or folded over the body, with the upperside visible.

Hesperiidae/Skippers

Regarded as the most primitive of butterflies, the skippers share the characteristics of both butterflies and moths. They are small and lively, with broad hairy bodies. Their eyes are widely separated and there is a bristle at the base of each antenna that is not found in other butterflies. The orange-chestnut skippers (subfamily Hesperiinae) sit in a characteristic pose with the forewings and hindwings held at different angles. Skipper larvae live in a shelter made from silk among their foodplants.

Papilionidae/Swallowtails

This family of spectacular insects contains the largest butterflies in the world, including the apollos, festoons, swallowtails and birdwings of Australasia and the Far East. In Britain the family is represented by just one species, the Swallowtail, although the Scarce Swallowtail may occasionally visit British shores. Swallowtail larvae are brightly coloured and have a forked organ – called the osmeterium – behind the head that emits a pungent smell if the larva is disturbed.

Pieridae/Whites & Yellows

Butterflies in this family have predominantly white or yellow wings with black markings. They include the garden whites, brimstones, orange tips and clouded yellows, and their colours usually derive from the pigments in their larval foodplants. Their larvae are long, slender and blunt at either end, and often have a decorated lateral line.

Lycaenidae/Hairstreaks, Coppers & Blues

About a third of the world's butterflies belong to this family. All species are relatively small and active. They have small slug-like caterpillars, and many have an intimate relationship with ants, whereby the ants attend the larvae, cleaning them, in return obtaining a sugary secretion from them. The *hairstreaks* have zigzag markings on the underside, and usually short tails. The *coppers* are characterized by their fiery copper colour, although a few are mostly dark brown and some are suffused with purple. Both sexes in the *blues* may be brown; in some the female only is brown.

Riodinidae/Metalmarks

This family is regarded as a subfamily of the Lycaenidae by some authorities. It contains some magnificent and diverse butterflies, the most spectacular of which occur in Central America. The single European species – the Duke of Burgundy – is one of the least striking members of the family.

Nymphalidae/Emperors, Vanessids & Fritillaries

This large family contains some of the biggest and gaudiest butterflies in the world, in Europe including the Peacock, Red Admiral and Purple Emperor, as well as the somewhat less spectacular fritillaries. A characteristic of adults of these butterflies is that the first pair of legs is vestigial, and only the remaining two pairs of legs are functional.

Danaidae

This family is considered by some to be a subfamily of the Nymphalidae. Only one member, the spectacular North American Monarch, occurs as a migrant in Britain. It dwarfs every other species found in Britain, and is one of the world's great migratory butterflies. Poisons from milkweeds, the foodplants of the striking black, yellow and white larva, protect it from predators.

Satyridae/Browns

The majority of butterflies in this family, which contains nearly a third of European butterflies, are brown or yellowish-orange; the exception is the Marbled White, which is more like members of the Pieridae. Butterflies in this group include browns, graylings, ringlets, woods and walls. They are medium or small in size, with small eyespots at the outer margins of the wings. The larvae typically taper towards the tail, and are often striped from head to tail, with colouration that matches the grasses on which they feed.

MOTHS

Moths tend to have hairier and stouter bodies than butterflies, and larger scales on the wings, which makes them look more fluffy and dense. Most are nocturnal or

diurnal, although there are exceptions to this generalization. Nocturnal moths are usually dull coloured with obscuring patterns that help to camouflage them when they are at rest during the day.

Micropterigidae
This family of very small moths is believed to be the most primitive of the Lepidoptera. The adults usually fly during the day, and have functional mandibles (rare in the Lepidoptera), which they used to feed on pollen.

Tortricidae/Leaf Rollers
Commonly known as tortrix moths, members of this family typically rest with the wings folded back, producing a somewhat rounded profile. There are more than 6,300 species worldwide.

Pyralidae/Pyralid Moths
Moths in this family generally rest with their wings in a triangular shape and put the first pair of long legs in front, the two antennae at the top and pointed backwards. They are small in size and have relatively long legs that extend beyond the hindwings when resting. There are more than 25,000 species of Pyralidae worldwide.

Pterophoridae/Plume Moths
The plume moths have unusually modified wings. The forewings usually consist of two curved spars with bedraggled bristles trailing behind, while the hindwings have three spars. At rest the wings are extended laterally and narrowly rolled up, and the moths may resemble a piece of dried grass, enabling them to avoid predators.

Hepialidae/Swift Moths
This family of more than 500 species worldwide is considered to be very primitive, with structural differences to other moths including very short antennae and the lack of a functional proboscis. Their forewings and hindwings are similar in size, and are folded along the body when at rest.

Cossidae/Leopard & Goat Moths

These are generally large and sturdy moths that fold their wings along the body when at rest. They tend to be grey in colour, and may mimic twigs, bark or leaves. Most larvae are tree borers, and in some species they take up to three years to mature. There are about 700 species worldwide.

Zygaenidae/Forester & Burnet Moths

The majority of the 1,000 or so species of Zygaenidae are tropical, but the family is also reasonably well represented in temperate regions. The Zygaenidae are typically day flying, with a slow and fluttering flight. They often have prominent red or yellow spots, the bright colours acting as a warning to predators that they are distasteful, containing hydrogen cyanide in all stages of their life cycle. This is manufactured by the moths themselves, rather than being obtained from a foodplant as is commonly the case among many Lepidoptera.

Sesiidae/Clearwing Moths

In these moths the wings have hardly any of the scales usually present in moths and butterflies, leaving them transparent. Their bodies are often striped with yellow, and they resemble wasps or hornets, probably as a defence against predation. There are 1,370 species, most of them in the tropics.

Lasiocampidae/Eggar Moths

Members of this family are large bodied with broad wings. They are either diurnal or nocturnal. Females are generally larger and slower than males, but the sexes are similar in other ways. Eggar moths are typically brown or grey, and have hairy legs and bodies. There are more than 2,000 species worldwide.

Saturniidae/Emperor Moths

This family includes large moths such as giant silkmoths, emperor moths and royal moths. It has members in the tropics with wingspans of up to 30cm, making them the largest insects on Earth today. All saturniids have large and heavy bodies with hair-like scales, lobed wings, small heads and reduced mouthparts. They are sometimes brightly coloured and may have transluscent eyespots on their wings. Males generally have broader antennae than females. There are up to 1,300 species worldwide, only one of which, the Emperor Moth, occurs in Britain.

Endromidae

The Kentish Glory is the only representative of this family. It is an attractive day-flying species that lives in birch forests throughout Europe and is scarce in Britain.

Thyatiridae/Lutestring Moths

This family comprises some 200 species, with 10 occurring on the Continent and 9 in Britain. It is named for the fine lines that cross the wings of some species. The wings of these moths are held tent-wise, close to the body, when at rest. The thorax often has prominent tufts.

Geometridae/Geometer Moths

This is a very large moth family of at least 20,000 species, more than 300 of which appear in Britain alone. Many have slender abdomens and broad wings, which are usually held flat with the hindwings visible. The majority fly at night, and the antennae of the male are often feathered. The moths are well camouflaged, with wavy wing patterns, and tend to blend in with the background. Their family name derives from geometer ('earth-measurer'), referring to the method of locomotion of the larvae. A larva will clasp its point of attachment with its front legs and draw up its hind end, employing a 'looping' gait, creating the impression that it is measuring its journey.

Sphingidae/Hawkmoths

Comprising about 1,050 species, the hawkmoth family is best represented in the tropics. The moths are moderate to large, and can employ rapid and sustained flight. Some, like the Hummingbird Hawkmoth, hover in midair while feeding on flower nectar and may be mistaken for hummingbirds. Most species are crepuscular or nocturnal, but some fly during the day. At rest the wings are generally laid flat over the body and swept back into an arrowhead shape. Hawkmoths are relatively long-lived, living for up to 30 days. The larvae of most species have a 'horn' at the rear end.

Notodontidae/Prominent & Kitten Moths
Species in this family are usually heavy bodied and long winged, and their wings are held folded across the back when at rest. They are generally grey or brown, and many species have a tuft of hair on the trailing edge of the forewing, which protrudes upwards when at rest, hence their common name of 'prominents'. The common names of some species, such as 'Puss Moth', refer to their hairiness. The family comprises more than 2,500 species, 27 of which have been recorded in Britain.

Lymantriidae/Tussock Moths
Moths in this family of about 2,700 species worldwide are usually muted brown and grey in colour, with some species being white, and tend to be very hairy. Their wings are generally held tent-wise when at rest, with the legs prominently displayed. Some females are flightless, and some have reduced wings. The larvae are hairy and frequently have tufts of hair that breaks off easily, forming a means of defence throughout their life cycle.

Arctiidae/Footman Moths, Tiger Moths, Ermines & Allies
This family comprises about 11,000 species worldwide, with 32 having been recorded in Britain. It includes brightly coloured tiger moths, as well as rather duller footmen, and lichen and wasp moths. Many species have hairy caterpillars, or 'woolly bears'. Many also retain distasteful or toxic chemicals that the larvae or adults acquire from foodplants, and advertise their defences with bright colouration,

scents or ultrasonic vibrations. Tiger moths hold their wings tent-wise when they are at rest, but most footmen lay them flat, with a considerable amount of overlap.

Thaumetopoeidae/Processionary Moths

The larvae of this small family of moths move in columns when searching for food, resembling a procession, hence the common name of the family. The adults are stout and furry, have feathered antennae and hold their wings close to their body at rest. Only two of about a hundred species worldwide have been recorded in Britain. These are the Pine Processionary and the Oak Processionary, the larvae of both of which may cause skin irritation in humans and can cause extensive defoliation of trees. Neither species is currently resident on mainland Britain, except as individuals, but they are expanding their range northwards, probably due to global warming. The Oak Processionary is resident on Jersey, where it was first recorded in 1984. Populations in southern Europe are controlled by natural predators that do not exist in northern Europe.

Noctuidae/Noctuid Moths

This is the largest of the moth families, with just over 400 species on the British list and about 21,000 species worldwide. Most British noctuids are medium-sized brown moths with a stout build and forewings considerably longer than they are deep, and most fly mainly at night. When at rest, the majority of noctuids hold their wings tent-wise over their body, with the trailing edges of the forewings brought together or slightly overlapping.

The 152 species in this book include all butterfly species and many of the most common moths currently found in Britain (including rarities like the Scarce Swallowtail and Monarch), and give an at-a-glance introduction to the butterflies and moths you are likely to see in the field. Measurements provided indicate the average size of a forewing; ranges are given for species that are particularly variable in size.

BUTTERFLIES

Grizzled Skipper
Pyrgus malvae

Male

Female

Male

SIZE AND DESCRIPTION
Forewing 12mm. Smaller than most skippers and with more white markings on a brown background. Sexes are similar, although male has more angular wings than female. Larva is green suffused with pale brown on the back, which is striped with darker olive-brown.

HABITAT AND DISTRIBUTION Open country, on grassy banks and at edges of woodland, from the lowlands to 1,800m. Europe except northern Britain and northern Scandinavia.

FOOD AND HABITS Flies April–August, in 1–2 broods. Larvae feed on Rock-rose, Wild Strawberry, Bramble and cinquefoils.

Dingy Skipper
Erynnis tages

Male

Female

Male

SIZE AND DESCRIPTION Forewing 14mm. Upperside is brown with tiny white spots near the margin of the forewing and hindwing. Underside is paler. Female is similar to male. Larva is green with a dark green line down its back, and a black head.

HABITAT AND DISTRIBUTION Usually banks of wild flowers on lime soils at up to 1,800m. Southern and central Europe, including England, Wales and southern Scandinavia.

FOOD AND HABITS Flies May–June, in 1–2 broods. Larvae feed on Bird's-foot Trefoil, Scorpion Vetch and other vetches.

Chequered Skipper
Caterocephalus palaemon

Female

Female

Male

Size and description Forewing 14mm. Upperside is dark brown with yellow patches. Underside is brownish with yellow scales and pale yellow spots on the hindwing. Female has more rounded wings and larger, more open markings than male. Larva is green with dark green and white lines, and a large green head.

Habitat and distribution Light woodland, rides and clearings from sea level to more than 1,000m in Alps. Most of France, western Scotland, central Europe and northern Scandinavia.

Food and habits Flies May–June. Larvae feed on grasses.

Large Chequered Skipper
Heteropterus morpheus

Male

Female

Male

SIZE AND DESCRIPTION Forewing 12mm. Dark brown with yellow markings. Sexes are similar, but female has more pronounced markings on the forewing than male, and chequered fringes. Larva is pale green.

HABITAT AND DISTRIBUTION Damp grasslands and woodland clearings. Mainly eastern Europe, but also northern Spain and western France. In Britain, believed to have been accidentally introduced to Jersey during the Second World War, but not seen on the island since 1996.

FOOD AND HABITS Flies June–July. Larvae feed on grasses and reeds.

Lulworth Skipper
Thymelicus action

Male

Female

Male

SIZE AND DESCRIPTION Forewing 12mm. Sexes are dissimilar and variable. Female is larger than male, with a ring of pale orange spots on the forewing. Male has a prominent black streak on forewing. Underside is dull orange in both sexes. Larva is green with a pale-bordered dark green line at the centre of the back, and pale greenish-yellow lines on either side.

HABITAT AND DISTRIBUTION Slopes of long grass, cliff tops and mountainsides. Most of Europe apart from north-east and Scandinavia; in Britain mainly along Dorset coast.

FOOD AND HABITS Flies May–July. Larvae feed on grasses.

Small Skipper
Thymelicus flavus

SIZE AND DESCRIPTION Forewing 14mm. Bright orange wings. Body is stout and rather moth-like. Tends to hold its wings flat when at rest. Has a swift darting flight.

HABITAT AND DISTRIBUTION Grassy places. England, Wales and mainland Europe south from Denmark.

FOOD AND HABITS Flies May–August. Small green larvae feed briefly on grasses, but go into hibernation shortly after hatching.

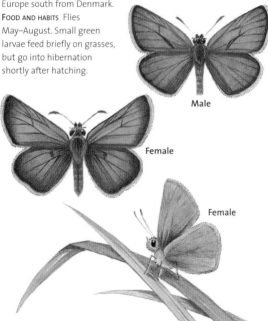

Male

Female

Female

Essex Skipper
Thymelicus lineola

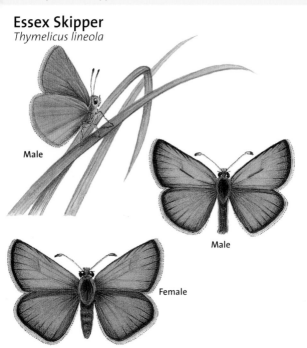

Male

Male

Female

SIZE AND DESCRIPTION Forewing 12mm. Very similar to Small Skipper (*T. flavus*), except for the black spots on the undersides of the antennae, and generally smaller. Male can be distinguished from female by the sex brand on his forewings.

HABITAT AND DISTRIBUTION Grassy places. Throughout Europe except northern Britain and northern Scandinavia.

FOOD AND HABITS Flies May–August. Larvae feed on grasses.

Silver-spotted Skipper
Hesperia comma

SIZE AND DESCRIPTION Forewing 15mm. Similar to Large Skipper (*Ochlodes venatus*), but the underside is spotted with white on a green background, the upperside is darker and less orange, and the pale yellow spotting is more prominent. Larva is dull olive-green with a black collar behind the head, which is large, and black marked with brown.
HABITAT AND DISTRIBUTION Grassy banks, meadows and cliffs. All of Europe apart from northern Britain and northern Scandinavia.
FOOD AND HABITS Flies July–August. Larvae feed on grasses.

Male

Female

Male

Large Skipper
Ochlodes venatus

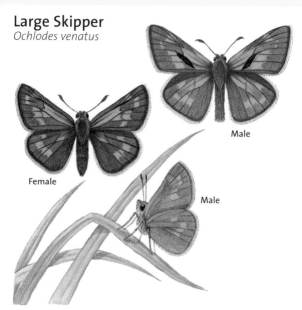

Male

Female

Male

Size and description Forewing 16mm. Upperside is orange-brown with dark veins and dark margins. Underside is paler with similar markings. Larva is blue-green with a dark line along its back and a yellow line down the sides.

Habitat and distribution Meadows, grassy banks and woodland edges, at up to 1,800m. Throughout Europe including Britain (absent in Ireland), to southern Scandinavia.

Food and habits Flies June–August. Only one brood in Britain and other northern areas, 1–3 elsewhere. Larvae feed on grasses such as Cock's-Foot. Overwinters as a larva.

Scarce Swallowtail
Iphiclides podalirius

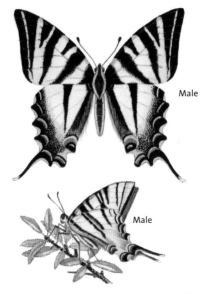

Male

Male

SIZE AND DESCRIPTION Forewing 40mm in male. Female is a little larger.
Sexes are similar – very pale creamy yellow with black markings, and
six stripes on the upper forewing. Larva looks like a green slug with
faint yellow stripes.

HABITAT AND DISTRIBUTION Often in fruit orchards at up to 1,800m.
Southern and eastern Europe. Vagrants occur in Britain very rarely.

FOOD AND HABITS Flies March–September. Two broods a year. Larvae feed
on Blackthorn and fruit trees.

Swallowtail
Papilio machaon

Size and description Forewing 38mm. Male and female are similar. Upperside is yellow with black markings, a red eyespot at the corner of each hindwing and a black band dusted with blue. Larva is up to 41mm long, and pale green striped black with red spots. When alarmed it flicks out two 'horns' – the osmeterium – on its head and emits a strong-smelling substance that deters predators.

Habitat and distribution Meadows and banks with wild flowers, especially umbellifers. Widespread in Europe; restricted to Norfolk in Britain, and occasional vagrants on south coast.

Food and habits Flies April–May and July–August. Has 2–3 broods in southern Europe. Larvae feed on umbellifers such as Milk Parsley, Fennel and Wild Carrot.

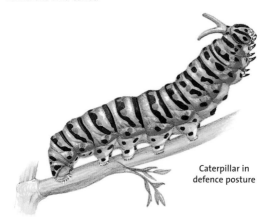

Caterpillar in defence posture

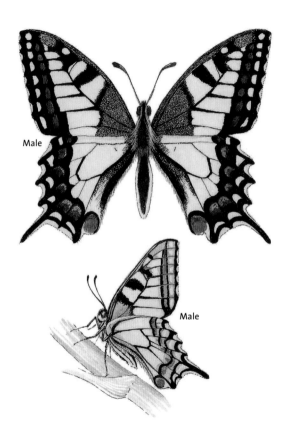

Male

Male

Small White
Artogeia rapae

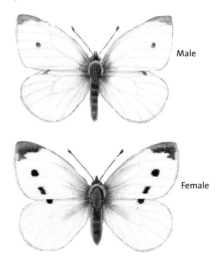

Male

Female

SIZE AND DESCRIPTION Forewing 25mm. Upperside is white with one black or grey spot on male's forewing and two on female's. Black or grey forewing patch extends further along the leading edge than down the side of the wing. There are two spots on the underside of the forewing in both sexes, and the underside of the hindwing is yellowish. Larva is mainly green.

HABITAT AND DISTRIBUTION Gardens, hedges and flowery places throughout Europe.

FOOD AND HABITS Flies March–October, in 2–4 broods. Larvae feed on brassicas and nasturtiums.

Large White
Pieris brassicae

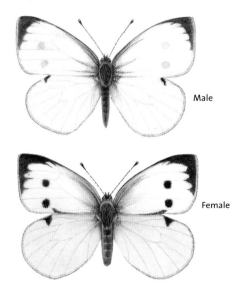

Male

Female

SIZE AND DESCRIPTION Forewing 30mm. Black tips extend halfway down the forewing's edge. Upperside of the forewing has two black spots in female. Underside of the forewing has two spots in both sexes. Larva is green with black spots and yellow stripes.

HABITAT AND DISTRIBUTION Gardens and other places with flowers throughout Europe.

FOOD AND HABITS Flies April–October, in 1–3 broods. Larvae feed on brassicas and nasturtiums.

Green-veined White
Artogeia napi

Female

Male

Female

SIZE AND DESCRIPTION Forewing 23mm. Black spots and patches on the forewing are less distinct than in Small and Large Whites (*A. rapae* and *Pieris brassicae*). There are grey lines along the veins on the underside of the hindwing. Larva is similar to that of Small White.

HABITAT AND DISTRIBUTION Gardens, hedges, woodland margins and other flowery places throughout Europe.

FOOD AND HABITS Flies March–November, in 1–3 broods. Larvae eat crucifers such as Garlic Mustard, Lady's Smock and Water-cress.

Black-veined White
Aporia crataegi

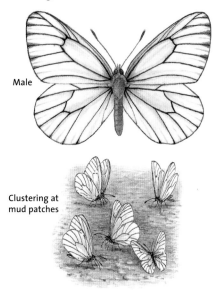

Male

Clustering at
mud patches

SIZE AND DESCRIPTION Forewing 30mm. Upperside is white, grey at tip of
forewing, with dark brown or black veins. Underside is similar with a
few black scales. Larva is grey with darker and red-brown lines.

HABITAT AND DISTRIBUTION Open country from sea level up to 1,800m.
Sometimes seen in clusters at wet mud patches. Widespread in
southern and central Europe; extinct in Britain since the 1920s.

FOOD AND HABITS Flies May–July. One brood a year. Larvae usually feed
on Hawthorn, but also on *Prunus* spp.

Orange Tip
Anthocaris cardamines

SIZE AND DESCRIPTION Forewing 23mm. Male has orange wing-tips and green blotches on the underside of the hindwing. Female has greyish patches on the forewing, and mottled underwings. Larva is up to 33m long, and green finely spotted with black.

HABITAT AND DISTRIBUTION Hedgerows, gardens, damp meadows and woodland margins. All Europe except south-west and east Spain, and northern Scandinavia.

FOOD AND HABITS Flies April–June. Larvae eat Garlic Mustard, Lady's Smock, and also Sweet Rocket and Honesty in gardens. Overwinters as a pupa.

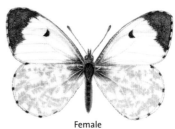

Female

Male

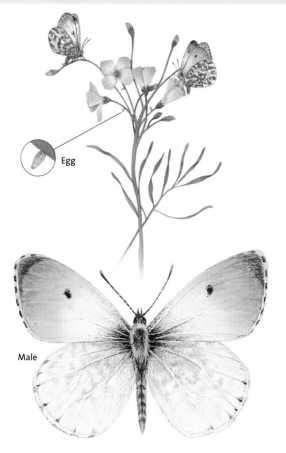

Egg

Male

Clouded Yellow
Colias crocea

Male

Female

Male

SIZE AND DESCRIPTION Forewing 25mm. Upperside is strong orange-yellow with black borders to both forewings and hindwings. Female has yellow spotting in black wing border. Underside of both sexes has black spots on a ground colour of dusky yellow-grey. Larva is green with a pale stripe.

HABITAT AND DISTRIBUTION Heaths and open areas at up to 1,800m. Southern and central Europe. Migrates to Britain.

FOOD AND HABITS Flies April–May onwards. Several broods a year. Larvae feed on clovers and vetches.

Pale Clouded Yellow
Colias hyale

Male

Female

Size and description Forewing 23mm. Male has a pale lemon yellow upperwing with dark grey or black borders. Female is white tinged with yellow-green. Larva is green and speckled black with a white stripe along each side.

Habitat and distribution Flowery meadows and fields at up to and above 1,800m. Southern and eastern Europe. Absent in Italy. Migrates north; rarely to Britain. May be mistaken for Berger's Clouded Yellow (*C. alfacariensis*), another rare migrant to Britain.

Food and habits Flies May–June and August–September. Two broods a year. Larvae feed on Lucerne and vetches.

Brimstone
Gonepteryx rhamni

SIZE AND DESCRIPTION Forewing 30mm. Male's wings are sulphur yellow on top, paler beneath. Female is almost white with a pale green tinge, but lacks the Large White's (*Pieris brassicae*) black markings. Larva is green with white stripes along the sides.

HABITAT AND DISTRIBUTION Open woodland, gardens and flowery places. All Europe, but not most of Scotland and northern Scandinavia.

FOOD AND HABITS Flies February–September. Larvae eat Buckthorn and Alder Buckthorn. Adults over-winter in hollies and ivies.

Caterpillar

Egg

Male

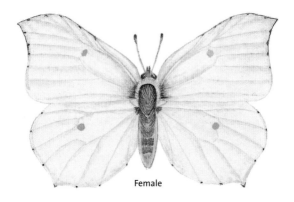

Female

Wood White
Leptidea sinapis

SIZE AND DESCRIPTION
Forewing 22mm. A delicate
white butterfly. Wings are
more oval than those of
other whites. Male has
rather darker charcoal
wing-tips than female.
Larva is green with a black
line along the centre of
its back.

HABITAT AND DISTRIBUTION
Woodlands and adjoining
meadows. Most of Europe,
but not northern Britain.
The similar Réal's White
(*L. reali*) is found in Ireland.
It has more slender wings,
darker wing-tips and
darker hindwings.

FOOD AND HABITS Flies
May–August. Usually
single brood a year in
Britain, 1–3 broods
elsewhere. Larvae feed on
vetches and related plants.

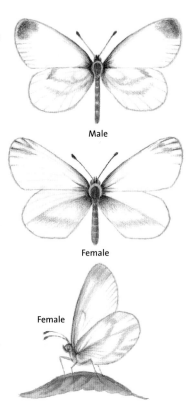

Male

Female

Female

Bath White
Pontia daplidice

Female

Male

Female

SIZE AND DESCRIPTION Forewing 22mm. Upperside is white with dark grey/black markings to the tip of the forewing; hindwing is greyer. Underside has grey-green markings. Female has larger markings on the forewing than male, and a dark spot on the underside of the forewing. Larva is greenish with three yellow stripes.

HABITAT AND DESCRIPTION Lowlands at up to 1,800m, on rough ground and meadows. Southern and central Europe; rare vagrant in Britain.

FOOD AND HABITS Flies February–March onwards. Migratory. Two or more broods a year. Larvae feed on a wide variety of crucifers.

Black Hairstreak
Strymonidia pruni

Female

Male

Male

SIZE AND DESCRIPTION Forewing 16mm. Upperside is dull brown with an orange border, which may be absent from the forewing. Black spots border both sides of the orange band on the underside of the hindwing. Larva is pale green; the head is withdrawn into its body.
HABITAT AND DISTRIBUTION Open woodland and hedgerows. Most of Europe except far north and south; in Britain confined to Midlands.
FOOD AND HABITS Flies July. Larvae feed on Blackthorn.

Green Hairstreak
Callophrys rubi

SIZE AND DESCRIPTION Forewing 15mm. Upperside is brown or grey; underside is green. Eyespots have white borders. Female is similar to male. Larva is green, broad and plump.

HABITAT AND DISTRIBUTION Heathland, moorland and rough ground with Heather or Gorse, at up to 2,100m. Widespread and common across Europe.

FOOD AND HABITS Flies March. One brood a year. Larvae feed on Bird's-foot Trefoil, Buckthorn, Bramble and gorses. Pupae overwinter and adults emerge in spring.

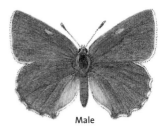

Male

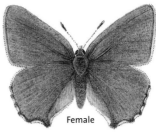

Female

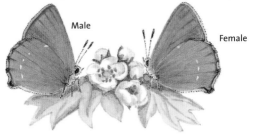

Male

Female

Brown Hairstreak
Thecla betulae

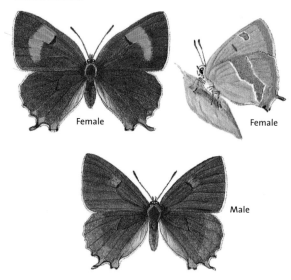

Female

Female

Male

SIZE AND DESCRIPTION Forewing 20mm. Upperside is brown with an orange patch on the forewing. Female is larger than male, with bright orange patches on the forewings. Underside is orange-yellow with white lines. Larva is plump and green.

HABITAT AND DISTRIBUTION Woodlands from low to moderate heights. Across central Europe including some parts of southern Britain.

FOOD AND HABITS Flies July–August. One brood a year. Eggs laid on twigs to hatch the following spring. Larvae feed on Blackthorn and other *Prunus* spp. leaves, and birches and beeches.

Purple Hairstreak
Quercusia quercus

Male

Male

Female

SIZE AND DESCRIPTION Forewing 19mm. Upperside is dark with a purplish-blue sheen and black margins in male, and bright purple forewing patches in female. Underside is grey with a white line and pale orange spots on the hind-wing corner. Larva is reddish-brown with a grey-brown line down its back.

HABITAT AND DISTRIBUTION Woodlands at up to 1,500m. Europe including Britain, but not northern Scandinavia.

FOOD AND HABITS Flies July–August. One brood a year. Larvae feed on oak buds and leaves. Eggs hatch the following spring.

White-letter Hairstreak
Strymonidia w-album

SIZE AND DESCRIPTION Forewing 16mm. Upperside is dark brown, sometimes with an orange flush on the forewing. Short tails to the hindwing. Pronounced white line on the hindwing, forming a letter 'W'. Female slightly larger than male, and not as dark brown. Larva is yellow-green, short and broad.

HABITAT AND DISTRIBUTION Woodlands and trees at up to 1,300m. Northern Spain to southern Scandinavia, Britain and Turkey.

FOOD AND HABITS Flies July. One brood a year. Larvae feed on leaves of Wych Elm and other elms. Eggs are laid in July–August to hatch the following spring.

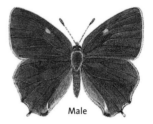

Male

Female

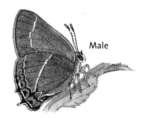

Male

Small Copper
Lycaena phlaeas

SIZE AND DESCRIPTION Forewing
15mm. Bright forewing is like
shiny copper, with dark flecks
and brown edges. Male is
smaller than female, and has
more pointed forewings. Larva
is small and green.

HABITAT AND DISTRIBUTION
Gardens, flowery wasteland
and heathland across Europe.

FOOD AND HABITS Flies
February–November. Two or
three broods a year; adults
from third brood rather small.
Larvae feed on Common and
Sheep's Sorrel, and docks.

Male

Female

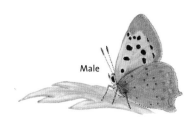

Male

Large Copper
Lycaena dispar

SIZE AND DESCRIPTION Forewing 20mm. Upperside of male is a brilliant coppery colour. Larger female is duller in colour, with mainly brown hindwings and a row of black spots on the forewing. Underside of the hindwing is silvery-grey spotted with black in both sexes. Larva is small and green.

HABITAT AND DISTRIBUTION Wet places such as fens in central and south-eastern Europe; endangered due to land drainage. Extinct in Britain since 1851, though reintroduction has been attempted.

FOOD AND HABITS Flies May–September, in 1–2 broods. Larvae feed on various dock species.

Female

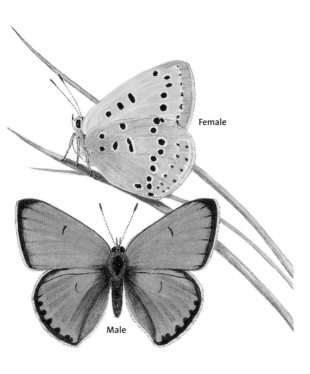

Female

Male

Short-tailed Blue
Everes argiades

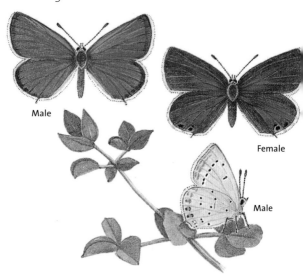

Male

Female

Male

SIZE AND DESCRIPTION Forewing 14mm. Male is violet-blue. In female the purple colouration is variable and almost absent in some individuals. A short tail on the hindwing and twin orange spots on the underside distinguish this species from other blues. Larva is pale green and slug-like. Also called Bloxworth Blue.

HABITAT AND DISTRIBUTION Damp flowery grassland. Southern and central Europe, and a very rare migrant to Britain.

FOOD AND HABITS Flies April–October, in 2–3 broods. Larvae feed on trefoils, vetches, gorses and allied plants.

Long-tailed Blue
Lampides boeticus

Male

Male

Female

SIZE AND DESCRIPTION Forewing 18mm. Upperside is violet-blue in male, sooty-brown with a variable blue flush in female. Underside is pale brown with paler markings. Both sexes have a tail on the hindwing. Larva is small and green.

HABITAT AND DISTRIBUTION Rough flowery places, meadows and among flowering crops. Strong migrant originating in southern Europe, North Africa and Asia. Migrates north, rarely into Britain and further north.

FOOD AND HABITS Flies May–October, in 2–3 broods. Larvae feed on buds and seed-pods of gorses, lupins, Bladder Senna and other members of the pea family.

Small Blue
Cupido minimus

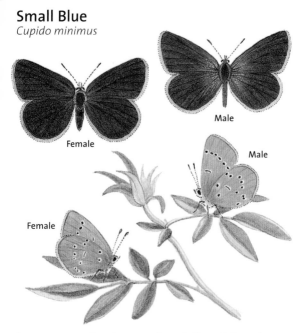

Female

Male

Male

Female

SIZE AND DESCRIPTION Forewing 12mm. Sexes similar sooty-brown in colour, but male has a dusting of gunmetal-blue scales, particularly in the basal area. Underside is powdery grey with black spots, and without any orange. Larva is brown and slug-like.

HABITAT AND DISTRIBUTION Flowery grassland, frequently on chalk and limestone. Europe except southern Spain and northern Scandinavia.

FOOD AND HABITS Flies June–July, in 1–2 broods. Larvae feed on Kidney Vetch.

Holly Blue
Celastrina argiolus

Male

First generation
female

Male

Second generation
female

SIZE AND DESCRIPTION Forewing 15mm. Upperside of male is violet-blue.
That of female is edged with a broad dark band, which is broader in
the second brood. Underside of the wings is pale blue-grey. Larva is
small, green and slug-like.

HABITAT AND DISTRIBUTION Woodland margins, hedgerows, parks and
gardens. Europe except Scotland and northern Scandinavia. The blue
most likely to be seen in gardens.

FOOD AND HABITS Flies April–September. Two broods a year. First brood
feeds on flowers and developing fruits of Holly; second on Ivy. Adults
drink honeydew, sap and the juices of carrion. Overwinters as a pupa.

Large Blue
Maculinea arion

Size and description Forewing 20mm. Upperside of female has a wider black border than that of male, and more prominent black spots. Underside is greyish-brown with a blue flush at the hindwing base and large black spots. Larva is pale cream and slug-like.
Habitat and distribution Areas with short grass and thyme. Most of Europe except far north. Declared extinct in Britain in 1979 and successfully reintroduced; some sites now contain large numbers.
Food and habits Flies June–July. Early larval foodplant is Wild Marjoram in western Europe, Wild Thyme in other areas. Older caterpillars feed on ant (*Myrmica* spp.) grubs – they are taken into the nest by ants, feed on the brood and pupate in spring.

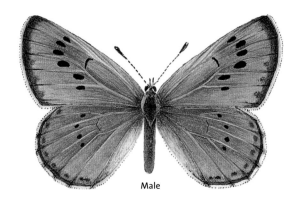

Male

Female

Female

Silver-studded Blue
Plebejus argus

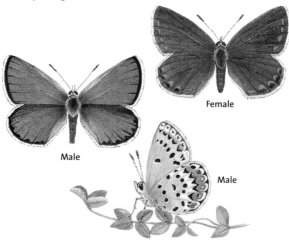

Male

Female

Male

SIZE AND DESCRIPTION Forewing 13mm. Upperside of male is deep blue with black borders. Female is usually brown. Underside is grey to greyish-brown, with dark white-ringed spots. Larva is green with a central dark black/brown stripe.

HABITAT AND DISTRIBUTION Grassy banks and heaths. Europe except northern England, northern Scandinavia, Ireland and the Mediterranean islands. Local in southern Britain.

FOOD AND HABITS Flies May. Two broods a year, but only one in northern Europe. Lays eggs in summer and larvae hatch next spring. Larvae feed on Bird's-foot Trefoil, heathers and gorses, and are tended by ants, which feed on their sticky secretions.

Brown Argus
Aricia agestis

SIZE AND DESCRIPTION Forewing 12mm. Upperside is chocolate-brown with no traces of blue, and the wings are bordered with orange spots. Female usually has larger spots than male. Underside is greyish-brown with orange spots along the margin. Larva is small and green, with a darker stripe along the centre back and a pale stripe along each side.

HABITAT AND DISTRIBUTION Downland, grassy slopes and open heaths. Europe except northern Britain and northern Scandinavia.

FOOD AND HABITS Flies May–August. Two broods a year. Larvae feed on Common Rock-rose, Common Stork's-bill and *Geranium* spp., and are attended by ants.

Male

Female

Female var.

Female

Northern Brown Argus
Aricia artaxerxes

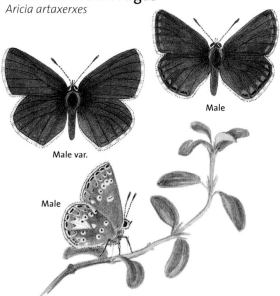

Male

Male var.

Male

SIZE AND DESCRIPTION Forewing 14mm. In Britain distinguished from Brown Argus (*A. agestis*) by the white spot on its forewing. In continental Europe the forewing tends to have little or no orange on the border. Larva is small and green.

HABITAT AND DISTRIBUTION Grassy slopes and sheltered moorland. At high levels over most of Europe; absent from Ireland and southern Britain.

FOOD AND HABITS Flies May–August. One brood a year. Larvae feed on Common Rock-rose and are attended by ants.

Chalkhill Blue
Lysandra coridon

Male

Male

Female

SIZE AND DESCRIPTION Forewing 18mm. Male is milky blue, female chocolate-brown. Underside is pale greyish-brown with black spots and faint orange markings. Larva is green with yellow stripes.
HABITAT AND DESCRIPTION Flowery areas on chalk and limestone. Southern and central Europe, to southern England.
FOOD AND HABITS Flies June–August. One brood a year. Larvae feed on Horseshoe Vetch and other vetches, and are attended by ants.

Adonis Blue
Lysandra bellargus

SIZE AND DESCRIPTION Forewing 17mm. Male upperside is a vivid iridescent blue, brighter than that of any other blue, with a black-spotted white fringe. Female is generally brown, often with varying amounts of blue. Larva is similar to that of Chalkhill Blue (*L. coridon*), but darker green.

HABITAT AND DISTRIBUTION Warm and sheltered spots on chalk downland. Southern and central Europe as far as southern England.

FOOD AND HABITS Flies May–June and July–August. Larvae feed on Horseshoe Vetch and have a particularly strong relationship with ants.

Female

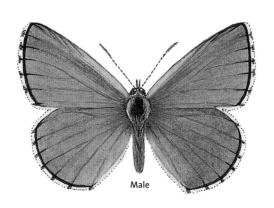

Male

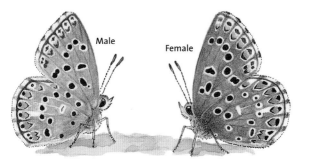

Male   Female

Common Blue
Polyommatus icarus

SIZE AND DESCRIPTION Forewing 15mm. Upperside of male's wings is violet-blue, that of female's wings dark brown. Larva is green and small.

HABITAT AND DISTRIBUTION Flowery grasslands, roadsides, sand dunes and wasteland across Europe.

FOOD AND HABITS Flies April–October, in 2–3 broods. Larvae feed on leguminous plants, particularly Bird's-foot Trefoil, and are generally less attractive to ants than other blues. Overwinters as a small larva.

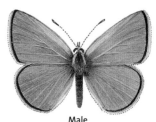

Male

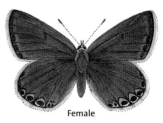

Female

Male

Female

Duke of Burgundy
Hamearis lucina

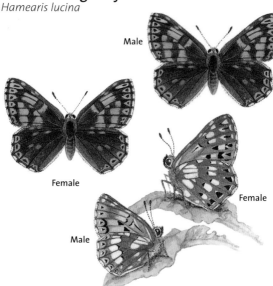

Male

Female

Female

Male

SIZE AND DESCRIPTION Forewing 15mm. Resembles a fritillary, so sometimes called the Duke of Burgundy Fritillary. Upperside of female is usually brighter than that of male. Both sexes have spots around all wing edges. Larva is brown and slug-like, but with long fine hairs.
HABITAT AND DISTRIBUTION Fields and open woodlands. Southern and central Europe, including southern Sweden; in Britain absent from Scotland, Wales and Ireland.
FOOD AND HABITS Flies May–August, in 1–2 broods. Larvae feed on Cowslip and Primrose.

Purple Emperor
Apatura iris

Male

Female

SIZE AND DESCRIPTION Forewing 36mm.
Upperside very dark, almost black, with
an iridescent blue flush, white spots
and stripes. Underside brown with
white markings. Orange eyespot under
forewing. Female is larger than male,
with bolder white markings and lacking
the blue flush. Larva is green with tiny
white spots, two horns on its head and
yellow stripes on its body.

Male

HABITAT AND DISTRIBUTION Woodland, in treetops, at up to 900m. Central
Europe including south-east England, where local.

FOOD AND HABITS Flies July–August. One brood a year. Larvae feed on
willow trees, especially Goat Willow and Grey Sallow. Lays eggs in
August, and overwinters as a larva.

Painted Lady
Cynthia cardui

Male

Male

SIZE AND DESCRIPTION Forewing 30mm. A fast-flying butterfly. Upperside is orange with a black forewing tip patched with white. Underside is paler. Black caterpillar has tufts of hairs and a yellow stripe on either side.

HABITAT AND DISTRIBUTION Flowery places including roadsides and gardens. Across Europe, but is a migrant from North Africa. Cannot currently survive British winters.

FOOD AND HABITS Flies April–November, arriving in Britain in late spring/early summer. Two broods a year in Europe, but broods throughout the year in North Africa. Larvae feed on thistles and sometimes Common Nettle.

Red Admiral
Vanessa atalanta

SIZE AND DESCRIPTION Forewing 30mm. Upperside is a velvety dark brown, with bright red bars on each wing. Tips of forewings are black with white markings. Underside of the hindwing is pale brown, while underside of the forewing shows red, blue and white markings. Dark larva has bristles and a pale yellow stripe along the sides.

HABITAT AND DISTRIBUTION Flowery places across Europe. Absent from northern Scandinavia. Resident in southern Europe, and moves north in spring.

FOOD AND HABITS Flies May–October. Two broods a year. Larvae feed on Common Nettle. Adults feed on rotting fruits in autumn.

Common Nettle, a larval foodplant

Male

Male

White Admiral
Limenitis camilla

Male

Female

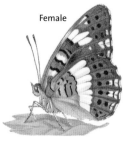

SIZE AND DESCRIPTION Forewing 28mm.
Upperside is dark with a broken line of
white spots on the forewing, and
a broad white line on the hindwing.
Underside is mottled brown and grey.
Sexes are similar in colour, but female
is slightly paler than male. Larva is
green with brown tufts down its back.
HABITAT AND DISTRIBUTION Lowland
woodland at up to 900m. Central Europe including southern England.
FOOD AND HABITS Flies June–July. One brood a year. Larvae feed on
honeysuckle and overwinter before pupating.

Large Tortoiseshell
Nymphalis polychloros

Male

SIZE AND DESCRIPTION Forewing 29mm.
Upperside is orange-brown with
dark markings, a dark border on the
forewing and a blue-spotted border
near the edge of the hindwing.
Sexes are similar. Underside is dark
and resembles a leaf – a form of
protective camouflage.

Male

HABITAT AND DISTRIBUTION Lowland
woodland at up to 1,500m in
southern and western Europe.
A very rare migrant to Britain.

FOOD AND HABITS Flies June–July. One brood a year. Larvae feed on trees
such as elms, willows, birches and poplars. Overwinters as adult.

Small Tortoiseshell
Aglais urticae

SIZE AND DESCRIPTION Forewing 24mm. Brighter than Large Tortoiseshell (*Nymphalis polychloros*). Upperside is orange and black, with a row of blue spots on the edges of the wings. Female similar to male, but has a fatter abdomen. Larva is up to 22mm long, yellow and black, and bristly.

HABITAT AND DISTRIBUTION All kinds of flowery places. Common across Europe, but has declined in Britain in recent years.

FOOD AND HABITS Flies March–October, in 2–3 broods. Adults overwinter, often in buildings. Larvae feed on nettles, buddleias, valerians and Michaelmas Daisies.

**Hibernating with
Peacock butterfly**

Male

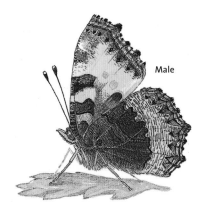

Male

Peacock
Inachis io

Size and description Forewing 29mm. Deep red wings have four large peacock-like 'eyes' – defensive markings that are a form of mimicry. Wings are flashed open when a predator approaches, with the forewings and hindwings simultaneously being rubbed together to produce a warning snake-like hissing sound. Underside is dark brown, resembling bark, and making the butterfly practically invisible on a tree-trunk. Caterpillar is up to 42mm long, and black and bristly.

Habitat and distribution Flowery places including gardens. Across Europe as far north as southern Scandinavia.

Food and habits Flies March–May and July–September. Males are territorial and may even challenge birds that fly over their territory. Larvae feed only on nettles. Adults often overwinter in buildings.

Caterpillar

Male

Male

Comma
Polygonia c-album

SIZE AND DESCRIPTION Forewing 25mm. Wings have jagged edges. Female's wings are less jagged and paler than male's. Upperside is orange with black and buff markings. Camouflaged underside of hindwing has a comma-shaped white mark. A bright golden form, called *hutchinsoni*, is produced by the earliest caterpillars, developing in spring. Larva is up to 35mm long, black and bristled. Its rear end becomes white, making it look like a bird dropping – a means of deterring predators.

HABITAT AND DISTRIBUTION Woodland margins, gardens, hedges and other flowery places. Across Europe, but absent from Ireland, northern Britain and northern Scandinavia.

FOOD AND HABITS Flies March–September. Two broods a year, with second brood darker than first. Larvae feed on nettles, hops and elms. Overwinters with adults hanging from leaves.

Caterpillar

Typical male

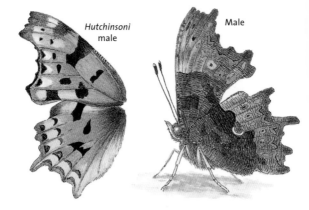

Hutchinsoni male

Male

Camberwell Beauty
Nymphalis antiopa

SIZE AND DESCRIPTION Forewing 33mm. Unmistakable upperside is sooty-brown with a pale border. Dark underwings provide camouflage when butterfly is hibernating in hollow trees. Attractive larva is up to 54mm long, and black and red with spines.

HABITAT AND DISTRIBUTION Woods and other places with trees. Widely distributed in Europe, but absent from Britain except as a rare vagrant; first discovered in Britain in Camberwell, London, hence its common name.

FOOD/HABITS Flies June–August and, after hibernation, March–April. Larvae feed on willows, sallows, birches and elms.

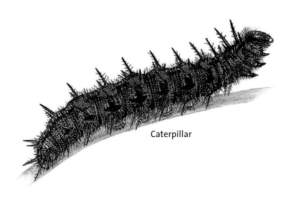

Caterpillar

Male

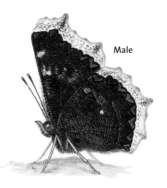

Male

Silver-washed Fritillary
Argynnis paphia

Male

Female

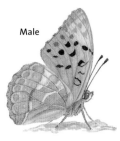

Male

SIZE AND DESCRIPTION Forewing 32mm. Upperside is tawny brown with black markings. Underside has a paler forewing, with a green-washed hindwing. Female is similar to male but rather duller. Larva is dark with spiky red-brown tufts.

HABITAT AND DISTRIBUTION Lowland woodland clearings at up to 1,400m. Southern and central Europe, including both southern England and southern Scandinavia.

FOOD AND HABITS Flies June–August. Single brood a year. Larvae feed mainly on Dog-violet. Overwinters as a larva.

Dark Green Fritillary
Mesoacidalia aglaia

Male

Male

SIZE AND DESCRIPTION Forewing 28mm. Upperside is tawny brown with dark markings. Underside forewing is yellow-buff with dark markings; hindwing is green-washed with white spots. Female is similar to male but paler. Larva is mainly dark brown with spiky black tufts.

HABITAT AND DISTRIBUTION Lowland meadows and heaths. Occurs throughout western Europe.

FOOD AND HABITS Flies June–July. One brood a year. Larvae feed on violets. Overwinters as a larva. Pupates after spinning a tent within leaves.

High Brown Fritillary
Fabriciana adippe

Female

Male

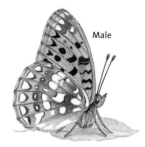

Male

SIZE AND DESCRIPTION Forewing
29mm. Strong-flying; female is
more heavily marked than male.
Larva is brown with a white line,
and reddish-brown or pink spines.
HABITAT AND DISTRIBUTION Flowery
meadows, woodland edges and
uplands. Central and southern
Europe. Once widespread in
England in Wales, but since the
1950s has undergone a dramatic decline; reduced to about 50 sites.
FOOD AND HABITS Flies July–August. Larvae feed on Dog Violet and
probably other violets.

Queen of Spain Fritillary
Issoria lathonia

Male

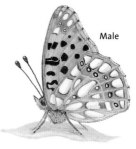

Male

SIZE AND DESCRIPTION Forewing 23mm. Medium-sized fritillary with large silver patches on the back-wing undersides. Upperside is a rich orange with black spots. Female is larger than male, with more green on the bases of the wings. Larva is mainly black, with a double white line along the middle of the back.

HABITAT AND DISTRIBUTION Flowery grasslands and edges of woodland. Throughout Europe, but a rare summer visitor to Britain.

FOOD AND HABITS Flies February–October. Two or more broods a year. Larvae feed on violets and closely related plants.

Pearl-bordered Fritillary
Clossiana euphrosyne

Male

Male

Female

SIZE AND DESCRIPTION Forewing 22mm. Small fritillary with light gold patches, two silver spots and a border of seven silver 'pearls' along the edges of the hindwing undersides. Upperside has the typical orange and black pattern of the fritillaries. Female is larger than male. Larva is black with spines.

HABITAT AND DISTRIBUTION Woodlands, particularly deciduous forests, and sometimes heathland. Found throughout Europe except Ireland and the Mediterranean coast.

FOOD AND HABITS Flies May–June, in 1–2 broods. Larvae feed on violets.

Small Pearl-bordered Fritillary
Clossiana selene

SIZE AND DESCRIPTION
Forewing 20mm. May
be confused with Pearl-
bordered Fritillary
(*C. euphrosyne*), but has
several silver spots on the
underside in addition to a
border of seven 'pearls'.
Employs a swift flapping
flight, often gliding low.
Larva is brown with two
horn-like bristles behind
the head.

HABITAT AND DISTRIBUTION
Open woodland in moist
areas across northern and
central Europe.

FOOD AND HABITS Flies
June–July and August–
September. Larvae feed on
Dog Violets, and hibernate
when they are less than
half grown.

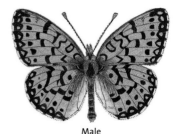

Male

Female

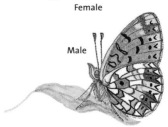

Male

Glanville Fritillary
Melitaea cinxia

Female

Male

Male

SIZE AND DESCRIPTION
Forewing 19mm, but
second brood is smaller.
Upperside is tawny brown
with dark patterning. Male
is slightly smaller than
female, with angular wings.
Underside forewing is mainly orange with a few dark markings, and
hindwing is patterned orange, black and cream. Larva is black with
white spots and short black tufts.

HABITAT AND DISTRIBUTION Lowland meadows at up to 1,800m. Southern
and central Europe, but only on Isle of Wight in Britain.

FOOD AND HABITS Flies May–June and August–September. Larvae feed
on Ribwort Plantain. Overwinters as a larva.

Heath Fritillary
Mellicta athalia

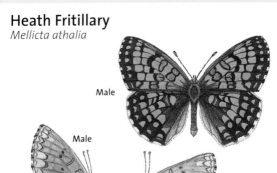

Male

Male

Female

Female

SIZE AND DESCRIPTION
Forewing 20mm. Darker
than Glanville Fritillary
(*M. cinxia*). Lack of dark
spots in submarginal bands
distinguish it from other
small fritillaries. Larva is black
spotted with greyish-white and has orange-yellow spines.

HABITAT AND DISTRIBUTION Flowery fields, open woods and woodland
edges. Widely distributed in Europe except southern Spain. Endangered
in Britain; found in only in a few woods in south and on Exmoor.

FOOD AND HABITS Flies May–September, in 1–3 broods. Larvae live
gregariously in a web, hibernating on Ribwort Plantain and Cow-
wheat, their main food sources.

Marsh Fritillary
Eurodryas aurinia

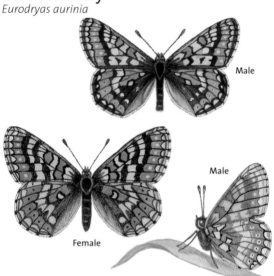

Male

Male

Female

SIZE AND DESCRIPTION Forewing 20mm. A bright orange, cream and white butterfly. Always a row of black spots on both sides of the hindwing, and no black spots on the underside of the forewing. Female is often larger than male. Larva is largely black with spines.

HABITAT AND DISTRIBUTION Marshy meadows, bogs, moors and boggy margins of lakes. Across Europe except far north. Endangered everywhere and rare in Britain.

FOOD AND HABITS Flies April–July. Larvae are gregarious and hibernate clustered at base of foodplant, Devil's-bit Scabious, and in continental Europe also Field Scabious, Greater Knapweed and plantains.

Monarch
Danaus plexippus

SIZE AND DESCRIPTION
Forewing 50mm.
Unmistakable world-
famous migrant
indigenous to North
America. Predominantly orange and black in colour, with male and
female being similar. Male has a small black pouch of scent scales on
a vein of the hindwing. Larva is strikingly coloured in yellow and black,
usually a deterrent to predators. Also called Milkweed.

HABITAT AND DISTRIBUTION Hot dry places. In Europe established only
in parts of the Canary Islands, southern Spain and Madeira. Adults
migrate, rarely coming to British shores.

FOOD AND HABITS Larvae feed on milkweeds. All stages in its life cycle
are poisonous due to toxins in the larval foodplant, to which larvae
are impervious but which are unpalatable to most birds.

Marbled White
Melanargia galathea

Male

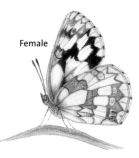

Female

SIZE AND DESCRIPTION Forewing 23mm. Upperwing is yellowish-white with very heavy black markings. Underside is paler, hindwing being grey and white with small eyespots. Hind underwing of female has a marked yellow tinge, as does the leading edge of the forewing. Larva is pale green or pale brown.

HABITAT AND DISTRIBUTION Grassy areas at up to 1,500m throughout the whole of western Europe, including southern England.

FOOD AND HABITS Flies June–July. Larvae feed on grasses such as Cock's-foot. Overwinters as a larva.

Grayling
Hipparchia semele

Female

Female

SIZE AND DESCRIPTION Forewing 24mm.
Upperside is brown washed with
orange, with two small eyespots
on the forewing and one on the
hindwing. Underside has a pale
orange forewing with two eyespots,
and a mottled hindwing. Upper wings
of female have broad yellow bands.
Larva is pale cream with darker stripes.
HABITAT AND DISTRIBUTION Heaths
and rough grassland with bare
ground patches for basking in
sunshine. Subspecies occur
throughout central and southern Europe; more coastal in Britain.
FOOD AND HABITS Flies July–August. Larvae feed on grasses.

Mountain Ringlet
Erebia epiphron

SIZE AND DESCRIPTION Forewing 19mm. Upperside velvety-brown with black-centred orange patches. Underside pale to deep brown with some orange on the forewing. Larva is small and green.

HABITAT AND DISTRIBUTION Craggy slopes at high altitudes. Mountainous habitats in southern and central Europe, and northern Britain.

FOOD AND HABITS Flies July–August. Larvae feed on grasses and hibernate.

Male

Female

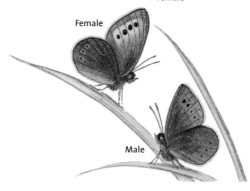

Female

Male

Scotch Argus
Erebia aethiops

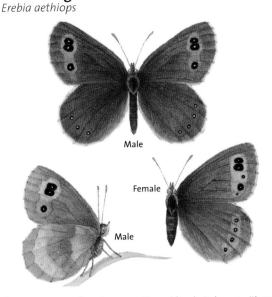

Male

Female

Male

SIZE AND DESCRIPTION Forewing 22mm. Upperside velvety-brown with an orange band containing white-centred eyespots. Underside similar to upperside, but the hindwing has 1–2 yellowish or silvery-grey bands, the outermost one with small eyespots. Female is generally paler than male. Larva is pale cream with pale and dark bands.

HABITAT AND DISTRIBUTION Found mainly on high ground. Central Europe and extreme northern Britain.

FOOD AND HABITS Flies July–September. Larvae feed on grasses and hibernate.

Meadow Brown
Maniola jurtina

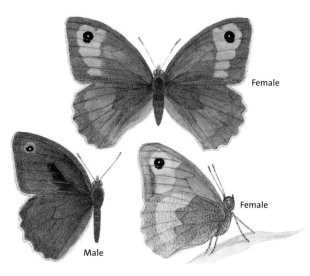

Female

Male

Female

SIZE AND DESCRIPTION Forewing 25mm. There is much variation in colour and size. Male typically lacks orange on the upper wings, which is present in female. Upper wing has a single black eye with a white highlight. Larva is green with a white stripe along the sides.

HABITAT AND DISTRIBUTION Grassland; also woodland in southern Europe. Very common across Europe southwards from southern Scandinavia at up to 2,000m.

FOOD AND HABITS Flies May–September. Larvae feed on grasses. Overwinters as a larva.

Ringlet
Aphantopus hyperantus

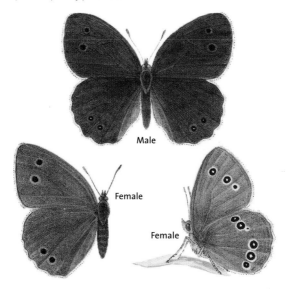

Male

Female

Female

SIZE AND DESCRIPTION Forewing 22mm. Upperside is very dark brown-black, with small and rather indistinct eyespots. Underside is paler with clearer eyespots ringed with yellow. Female is slightly paler than male. Larva is pale cream with a lighter band along the sides.
HABITAT AND DISTRIBUTION Woodland rides and clearings, and damp grassy areas at up to 1,500m in western Europe.
FOOD AND HABITS Flies June–July. Larvae feed on grasses. Overwinters as a larva, then pupates on ground.

Gatekeeper
Pyronia tithonus

Male

Female

Male

SIZE AND DESCRIPTION Forewing
22mm. Usually smaller than
Meadow Brown (*Maniola jurtina*), with orange patches on the wings.
'Eyes' are black with two highlights on each forewing. Male has a
broad band of dark scent scales on each forewing, and a fuller orange
colour than female, which is paler and larger. Larva is green or brown.
HABITAT AND DISTRIBUTION Hedgerows and woodland margins. Southern
Britain and Ireland, and south across rest of Europe.
FOOD AND HABITS Flies July–September. Larvae feed on fine-leaved
grasses. Adults are fond of Bramble blossoms and Wild Marjoram.

Large Heath
Coenonympha tullia

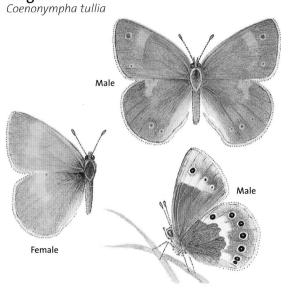

Male

Female

Male

SIZE AND DESCRIPTION Forewing 19mm. Upperside is dull grey-brown washed with orange; forewing has small eyespots and hindwing has several eyespots. Underside has golden-brown forewing, slightly darker hindwing and more distinctive eyespots. Female is usually slightly paler than male. Larva is green with yellow lines.

HABITAT AND DISTRIBUTION Flat wet areas such as bogs and waterlogged peat mosses. Northern and eastern Europe.

FOOD AND HABITS Flies June–early July. Larvae feed on Hare's-tail.

Small Heath
Coenonympha pamphilius

SIZE AND DESCRIPTION Forewing
16mm. Male is generally
smaller and brighter than
female. Colouration is highly
variable. Upperside is orange-
brown with very small dark
'eyes'. Underside hindwing is
greyish, and forewing has a
distinctive spot. Larva is
mainly green.

HABITAT AND DISTRIBUTION
Grassy places at up to
2,000m. Across Europe
except northernmost
Scandinavia.

FOOD AND HABITS Flies
April–October, in 1–3 broods.
Pupa feeds on grasses.
Overwinters as a larva.

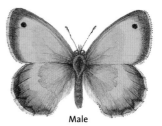

Male

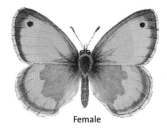

Female

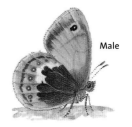

Male

Speckled Wood
Pararge aegeria

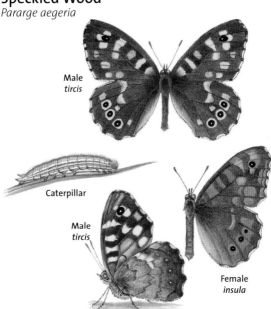

Male
tircis

Caterpillar

Male
tircis

Female
insula

Size and description Forewing 20mm. Yellow-and-brown or orange-and-brown wings. Orange-spotted form (*P. a. insula*) south-west Europe and Italy; cream-spotted form (*P. a. tircis*) elsewhere. Larva is mainly green and up to 27mm long.

Habitat and distribution Woodland clearings, gardens and paths across Europe from southern Scandinavia.

Food and habits Flies March–October. Single-brooded in north. Larvae feed on grasses. Overwinters in both larval and pupal forms.

Wall Brown
Lasiommata megera

Male

Female

Male

SIZE AND DESCRIPTION Forewing 22mm. Brown-and-orange patterned with an 'eye' on the forewing. Male typically has an oblique line of scent scales on the forewing. Underside of the hindwing is pale silvery-brown. Female is lighter and brighter than male. Larva is green.
HABITAT AND DISTRIBUTION Rough grassy places and gardens. Britain (except northern Scotland) and across from southern Europe.
FOOD/HABITS Flies March–October, in 2–3 broods. Adults sunbathe on walls and fences. Larvae feed on grasses. Overwinters as a larva.

MOTHS

Tapestry Moth
Trichophaga tapetzella

SIZE AND DESCRIPTION Forewing 9mm. Greyish-white wings have brown patches towards the thorax.

HABITAT AND DISTRIBUTION Stables and buildings with high humidity.

FOOD AND HABITS Flies June–July. A clothes moth whose larvae feed on animal fibres such as wool, and can damage textiles. Found in horsehair stuffing and owl pellets.

Leaf-mining Moth
Stigmella aurella

SIZE AND DESCRIPTION Forewing 3mm. Forewings have yellow bars and
purplish wingtips. Pale and feathery underwings. The tiny leaf-mining
larva creates pale squiggly lines on Bramble leaves.

HABITAT AND DISTRIBUTION Woodland, hedges and gardens over most of
Europe, except far north.

FOOD AND HABITS Flies May–September. Larvae feed on Bramble. They
overwinter in the leaf-mines, but leave before pupating.

Lampronia rubiella

SIZE AND DESCRIPTION Forewing 5mm. Yellow or cream bars on dark brown forewings. Underwings feathered at the lower edges.
HABITAT AND DISTRIBUTION Gardens with Raspberries across northern and central Europe.
FOOD AND HABITS Flies May–June. Larvae feed in the central stalks of Raspberry fruits in summer. They overwinter in the soil, then complete their growth in buds during spring.

Common Clothes Moth
Tineola bisielliella

SIZE AND DESCRIPTION Forewing 5mm. Goldish forewings and silvery hindwings. Larva is white with a pale brown head.

HABITAT AND DISTRIBUTION Rarely seen outdoors. The most common and destructive clothes moth.

FOOD AND HABITS Adults found throughout the year. Rarely flies, preferring to scuttle for cover. Larvae eat animal fibres and also build shelters out of them.

Green Oak Roller
Tortrix viridana

SIZE AND DESCRIPTION Forewing 10mm. Pale green forewings and pale grey hindwings. Green larva measures about 12mm in length.
HABITAT AND DISTRIBUTION Woods, parks and gardens with oaks.
FOOD AND HABITS Flies May–August at night, but lives for only one week. Larvae feed on the buds and rolled leaves of oak trees. Will hang on a thread from trees.

Codlin Moth
Cydia pomonella

SIZE AND DESCRIPTION Forewing 9mm. Grey forewing has black and yellowish marks towards the tip. White larva has a brown head, becoming pinkish as it grows larger.

HABITAT AND DISTRIBUTION Orchards, parks, gardens and hedges with apple trees. All Europe except far north.

FOOD AND HABITS Flies May–October. Two broods. Larvae bore into apples (and pears) to eat both the flesh and developing seeds. Pupates under loose bark in a cocoon, from which it emerges during spring.

Small Magpie
Eurrhypara hortulata

SIZE AND DESCRIPTION Forewing 15mm. Silky white with dark grey markings and a yellowish-gold thorax with black spots. Larva is green.
HABITAT AND DISTRIBUTION Hedgerows, woodland margins and waste ground with nettles. All Europe except far north.
FOOD AND HABITS Flies June–August. Larvae feed on Common Nettle and related plants. Overwinters as a larva in a spun cocoon located among plant debris.

Gold Fringe
Hypsopygia costalis

SIZE AND DESCRIPTION Forewing 8mm. Dark brown forewings have two gold marks and a golden-yellow fringe. Hindwings are purplish with a gold fringe. Larva is whitish with a brown head.

HABITAT AND DISTRIBUTION Hedges around grassy places in southern Britain and southern and central Europe.

FOOD AND HABITS Flies July–October. Larvae feed on dead grasses and thatch.

White Plume
Pterophorus pentadactyla

SIZE AND DESCRIPTION Forewing 13mm. White with each forewing being split into two feathery sections and each hindwing into three feathery sections. Bright green larva has tufts of silvery hair.

HABITAT AND DISTRIBUTION Hedgerows, waste ground and gardens throughout Europe.

FOOD AND HABITS Flies May–August at night. Larvae feed on bindweeds, curled up in a leaf. Overwinters as a small larva.

Ghost Swift
Hepialus humuli

SIZE AND DESCRIPTION Forewing 21mm. Male (shown here) is a very pale creamy white. Female is a little larger and darker. Larva is very pale white to grey-white, with dark spots and a red-brown head.

HABITAT AND DISTRIBUTION Downland, meadows and gardens. Widespread throughout northern and central Europe, including Britain.

FOOD AND HABITS Flies June. Eggs laid in flight. Larvae ingest grasses and herbaceous plants, feeding below ground on roots. May take two years to reach pupation stage, which occurs in May.

Common Swift
Hepialus lupulinus

SIZE AND DESCRIPTION Forewing 16mm. Brown wings with white marks.
Very short antennae. Wings are held tightly to the body when at rest.
White or creamy larva is about 35mm long with a brown head.
HABITAT AND DISTRIBUTION Arable land, gardens, parks and grassland over
most of Europe, except Iberia.
FOOD AND HABITS Flies May–August at dusk. Larvae live in the soil,
eating the roots of grasses and other herbaceous plants. Overwinters
as a larva.

Goat Moth
Cossus cossus

SIZE AND DESCRIPTION Forewing 30mm. Greyish-brown wings with fine dark patterning on the forewings. Abdomen is distinctly ringed. Large and solid-looking. The purplish-red larva emits a goat-like smell.

HABITAT AND DISTRIBUTION Broadleaved woodland from Ireland and England south across Europe.

FOOD AND HABITS Flies June–August. Larvae feed on solid wood of broadleaved trees for approximately three years before pupating in the ground.

Leopard Moth
Zeuzera pyrina

SIZE AND DESCRIPTION Forewing 20–35mm. White with finely spotted wings and six black marks on the furry thorax. Abdomen is ringed with greyish black. Female is much larger than male. The creamy larva has black spots and a dark head.

HABITAT AND DISTRIBUTION Woods, parks, orchards and gardens. Found from England across central and southern Europe.

FOOD AND HABITS Flies June–August at night. Single-brooded. Larvae tunnel into broadleaved trees and shrubs.

Forester
Adscita statices

SIZE AND DESCRIPTION Forewing 13mm. Upperside forewing is metallic green. Larva is dull yellow with tiny black spots and small tufts.

HABITAT AND DISTRIBUTION Chalk downland, sea cliffs, meadows and heaths. Occurs widely throughout Europe.

FOOD AND HABITS Flies by day during June. One brood a year. Larvae feed on sorrels. Hibernates as a larva.

Five-spot Burnet
Zygaena trifolii

SIZE AND DESCRIPTION Forewing 16mm. Upperside has metallic green forewings with five crimson spots. Hindwing is crimson bordered with green. Larva is very like that of the Six-spot Burnet (*Z. filipendulae*), but more blue-green.

HABITAT AND DISTRIBUTION Chalk downland, meadows and marshy ground. Widespread in Europe, but in Britain found mainly in southern England and parts of Wales.

FOOD AND HABITS Flies May–July. One brood a year. Larvae feed on Bird's-foot Trefoil. Overwinters as a larva. May take two years to pupate. Flies by day.

Six-spot Burnet
Zygaena filipendulae

SIZE AND DESCRIPTION Forewing 18mm. Upperside has a very dark metallic green-black forewing with six crimson spots; hindwing is crimson bordered with green. Larva is greenish-yellow with two lines of large black spots along the back.

HABITAT AND DISTRIBUTION Meadows, downland, sea cliffs and woodland. Distributed widely throughout Europe and Britain, but is more coastal in Scotland.

FOOD AND HABITS Flies June–July. Larvae feed on Bird's-foot Trefoil and other vetches. Overwinters as a larva. May take two years to achieve maturity. Flies by day.

Currant Clearwing
Synanthedon tipuliformis

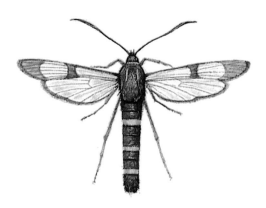

SIZE AND DESCRIPTION Forewing 8mm. Wings are largely transparent. Black abdomen has four yellow rings in male and three in female. The moth usually rests with its wings apart. The dingy white larva has a brown head and yellow spots.

HABITAT AND DISTRIBUTION Woods, gardens and open country with suitable foodplants across Europe.

FOOD AND HABITS Flies May–July. One brood a year. Larvae feed inside the stems of Blackcurrant, Whitecurrant and Gooseberry. Overwinters as a larva. Flies by day.

Large Red-belted Clearwing
Synanthedon culiciformis

SIZE AND DESCRIPTION Forewing 11–14mm. The red belt refers to a band on the abdomen. Wings are bordered with black.

HABITAT AND DISTRIBUTION Open woodland and heathland. Common and widespread in Europe (but not found in Ireland).

FOOD AND HABITS Flies May–June. One brood a year. Larvae feed on birches under the bark of stumps, sometimes in alders, where they pupate. Overwinters as a pupa. Flies by day.

Lackey
Malacosoma neustria

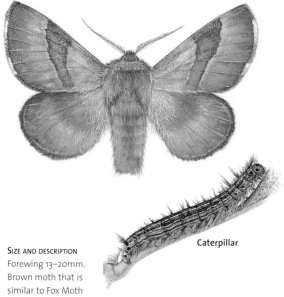

Caterpillar

SIZE AND DESCRIPTION
Forewing 13–20mm.
Brown moth that is
similar to Fox Moth
(*Macrothylacia rubi*), but wing bands curve inwards. Tufted grey-blue
larva has white, orange, black and yellow stripes along its body.

HABITAT AND DISTRIBUTION Many habitats over most of Europe, except
Scotland and northern Scandinavia.

FOOD AND HABITS Flies June–August at night. One brood a year. Larvae
live in colonies in cocoons, feeding on the leaves of Hawthorn,
Blackthorn, plums and sallows. Overwinters as an egg.

Oak Eggar
Lasiocampa quercus

SIZE AND DESCRIPTION Forewing 29–45mm. Upperside of both forewings and hindwings is rich brown, the outer third being paler. Small and dark-ringed white spot on each forewing. Female considerably larger than male (shown here), and paler. Larva is hairy and dark brown with black rings.

HABITAT AND DISTRIBUTION Woodland and hedgerows, as well as heathland and moors. Throughout Europe and widespread in Britain.

FOOD AND HABITS Flies May–June. Male flies by day, female by night. Larvae feed on wide range of plants, depending on subspecies, including oaks, heathers and Bramble. Overwinters as a larva.

Lappet
Gastropacha quercifolia

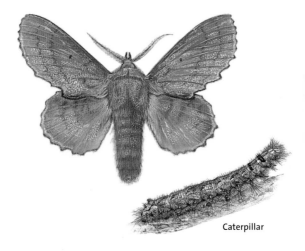

Caterpillar

SIZE AND DESCRIPTION Forewing 28–42mm. Female much larger than male. Colour varies across its range, from purple in north to pale brown in south. At rest the wings resemble dead leaves. Larva is up to 80mm long, and dark brownish-grey with two bluish bands near the head. Row of large fleshy flaps along the sides of its body helps it to blend in against the foliage on which it rests.

HABITAT AND DISTRIBUTION Open woodland, hedges, orchards and gardens. Most of Europe except Scotland, Ireland and much of north.

FOOD AND HABITS Flies May–August at night. One brood a year. Larvae feed on Blackthorn, Buckthorn, apples and sallows. Overwinters as a larva.

Fox Moth
Macrothylacia rubi

SIZE AND DESCRIPTION Forewing 20–30mm. Male is fox-coloured with two narrow pale stripes on each forewing; female is paler. Larva is velvety and very dark brown, with orange bands.

HABITAT AND DISTRIBUTION Heathland, moorland, open countryside and woodland margins throughout Europe.

FOOD AND HABITS Flies May–July. Male flies in sunshine and at night; female is a purely nocturnal flier. Larvae feed on Bramble, Bilberry, Creeping Willow and heathers. Overwinters as a full-grown larva.

Drinker
Philudoria potatoria

SIZE AND DESCRIPTION Forewing 25–35mm. Male's (shown here) upperwing is orange-brown with two dark lines and two white spots. Female is slightly larger with similar wing markings but yellow-orange ground colour. Larva is hairy, blue-grey and has black sides.
HABITAT AND DISTRIBUTION Open woodland, damp moors and fens throughout continental Europe and Britain.
FOOD AND HABITS Flies July. One brood a year. Larvae feed on grasses such as Cock's-foot and Couch Grass. Overwinters as a larva.

Kentish Glory
Endromis versicolora

Size and description Forewing 28–36mm. Upperside forewing is buffy-brown with white and black markings. Hindwing is more yellow. Female larger and paler than male (shown here). Larva is large and green, with whitish oblique stripes and a hump on the rear of its body.

Habitat and distribution Woodland and moorland with woods. Widespread in continental Europe, but localized in Britain. Extinct in England, but still occurs in northern Scotland.

Food and habits Flies April. Larvae feed mainly on Silver Birch and sometimes on alders. Overwinters as a pupa. Mostly nocturnal, but male sometimes flies by day.

Emperor Moth
Saturnia pavonia

Caterpillar

SIZE AND DESCRIPTION Forewing 34–42mm. Large moth with single big and startling eyespots on both the forewings and hindwings. Female larger and paler than male (shown here). Larva is up to 60mm long, and green, banded black, with tufts.

HABITATS AND DISTRIBUTION Moors, heaths, mountainsides and woodland. Occurs widely throughout Europe, including Britain.

FOOD AND HABITS Flies in spring, laying eggs in May. Larvae feed principally on Heather but also on a range of other plants including Bramble, Blackthorn, Hawthorn and Bilberry. Overwinters as a pupa.

Peach Blossom
Thyatira batis

SIZE AND DESCRIPTION Forewing 15mm. Forewings are brown with pink blotches. Larva is dark brown with slanting white lines and bumps on its back.

HABITAT AND DISTRIBUTION Woodland and woodland edges in northern and central Europe, including Britain.

FOOD AND HABITS Flies May–August at night. One brood a year. Larvae feed on Bramble, Raspberry and Blackberry. Overwinters as a pupa.

Large Emerald
Geometra papilionaria

SIZE AND DESCRIPTION Forewing 25–32mm. Upperside is pale blue-green marked with faint white lines and spots. Sexes are similar, but shade of green and number of markings may vary. Larva is rough-skinned, body coloured yellowish-green with reddish warts on the back.

HABITAT AND DISTRIBUTION Woodland, moors, heathland and fens. Occurs widely in central and northern Europe, including Britain.

FOOD AND HABITS Flies July. Larvae feed on trees such as Silver Birch, alders, beeches and Hazel. Overwinters as a larva.

Winter Moth
Operophtera brumata

SIZE AND DESCRIPTION Forewing 15mm. Male (shown here) has greyish-brown wings with a faint pattern; female has stunted relict wings. The green looper caterpillar is about 20mm long.

HABITAT AND DISTRIBUTION Abundant throughout Europe wherever there are trees and shrubs.

FOOD AND HABITS Flies October–February. Nocturnal and attracted to lighted windows. Females can be seen on windowsills and tree-trunks. Larvae feed on deciduous trees.

Garden Carpet
Xanthorhoe fluctuata

SIZE AND DESCRIPTION Forewing 14mm. Greyish-white wings with markings in various shades of grey. Pattern varies, but there is always a dark triangle where the forewings join the thorax. Larva is a twig-like grey-green to dark brown looper.

HABITAT AND DISTRIBUTION Common in cultivated areas.

FOOD AND HABITS Flies April–October from dusk, in 2–3 broods. Rests on walls and fences during the day. Larvae feed on Garlic Mustard, Shepherd's-purse and other crucifers. Overwinters as a pupa.

Small Emerald
Hemistola chrysoprasaria

SIZE AND DESCRIPTION Forewing 18mm. Pale grey-green with two fine white lines on the forewing and a single line on the hindwing. Larva is pale green with white dots and a brown head.

HABITAT AND DISTRIBUTION Downland, hedges and woodland edges, usually on chalk or limestone.

FOOD AND HABITS Flies May–August at night. Single-brooded. Larvae feed on Traveller's Joy.

Brimstone Moth
Opisthograptis luteolata

SIZE AND DESCRIPTION Forewing 18mm. Sulphur-yellow with brown flecks on the leading edge of the forewing. Brown looper larva is about 30mm long and often tinged grey or green, with a prominent nodule on the middle of its back.

HABITAT AND DISTRIBUTION Woods, hedges and gardens across Europe.

FOOD AND HABITS Flies April–October at night. Larvae feed on Hawthorn, Blackthorn and other shrubs.

Lime-speck Pug
Eupithecia centaurearia

SIZE AND DESCRIPTION Forewing 12mm. Pale grey with dark marks on the forewings. Green or yellow larva often has red spots.
HABITAT AND DISTRIBUTION Rough areas and gardens throughout Europe.
FOOD AND HABITS Flies May–October from dusk. May be double-brooded. Rests with its wings outstretched on lichens on walls and tree-trunks. Larvae feed on a range of herbaceous plants such as Yarrow and ragworts. Overwinters as a pupa.

Magpie
Abraxas grossulariata

SIZE AND DESCRIPTION
Forewing 20mm. Variable
black-and-white pattern,
with a yellowish-orange
line across the middle of
the forewing and near the
head. Looper larva is up
to 32mm long, and pale
green with black spots and
a rusty line along its sides.

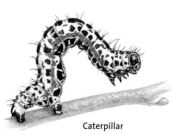

Caterpillar

HABITAT AND DISTRIBUTION Woods, gardens and hedges throughout
Europe except far north.

FOOD AND HABITS Flies June–August. Larvae feed on Blackthorn,
Hawthorn, Gooseberry and many other shrubs. Overwinters as a
small caterpillar and pupates in May or June.

August Thorn
Ennomos quercinaria

SIZE AND DESCRIPTION Forewing 17mm. Pale yellowish-tan with two narrow brown stripes on each forewing. Female paler than male (shown here). Abdomen is fluffy. Greyish-brown looper caterpillar has nodules that make it look like a twig.

HABITAT AND DISTRIBUTION Woodland, parks and gardens across Europe.

FOOD AND HABITS Flies August–September at night. Larvae feed on oaks and other trees.

Large Thorn
Ennomos autumnaria

Size and description Forewing 12–25mm. Upperside varies from pale to deep yellow with purplish-brown speckling. Larva looks like a little twig and is very slender, coloured brown or greenish-brown.
Habitat and distribution Woodland and bushy areas across Europe.
Food and habits Flies September. Larvae feed on trees such as Silver Birch, alders, Hawthorn and Blackthorn. Overwinters as an egg, which hatches in spring.

Swallow-tailed Moth
Ourapteryx sambucaria

SIZE AND DESCRIPTION Forewing 30mm. Wings are bright lemon, rapidly fading to pale cream or white. The pointed tail with two dark spots on the hindwing is a diagnostic feature. Larva is a brown looper measuring up to 50mm in length.

HABITAT AND DISTRIBUTION Forest edges, woods, gardens, scrub and parks across Europe except far north.

FOOD AND HABITS Flies June–August at night, often coming to lighted windows. Larvae feed on Blackthorn, Hawthorn, ivies, and numerous other trees and bushes.

Peppered Moth
Biston betularia

SIZE AND DESCRIPTION Forewing 20–30mm. Variable. Normal form is white peppered with fine dark marks; also sooty black. Green or brown looper larva is up to 60mm long.

HABITAT AND DISTRIBUTION Woods, gardens, scrub and parks across Europe except far north.

FOOD AND HABITS Flies May–August, coming to lighted windows. Larvae feed on a range of trees and shrubs, including sallows, Hawthorn, Golden Rod and Raspberry.

Mottled Umber
Erannis defoliaria

SIZE AND DESCRIPTION Forewing 18–25mm. Upperside of male (shown here) is creamy-brown, forewing with darker bands, hindwing pale and speckled. Female is wingless and clings to trunks and the stems of foodplants. Larva is dark olive-green with a greenish-yellow broken stripe along its sides.

HABITATS AND DISTRIBUTION Woodland throughout Europe; rare in Scotland and Ireland.

FOOD AND HABITS Flies mainly October–December. Larvae feed on a variety of trees, especially birches, oaks, hornbeams, Blackthorn and Hawthorn. Overwinters as an egg, which hatches in spring.

Poplar Hawkmoth
Laothoe populi

Caterpillar

SIZE AND DESCRIPTION
Forewing 30–46mm.
Variable colouring, from
grey to pinkish-brown.
Orange patches on the
hindwings. At rest gives
the impression of a
bunch of dead leaves. Green larva, up to 65mm long, has a yellow
horn and seven diagonal yellow stripes.

HABITAT AND DISTRIBUTION Woodland margins, river valleys and parks
throughout Europe except far north.

FOOD AND HABITS A slow-flying moth that is on the wing in May–
September. May have 1–2 broods a year. Larvae feed on poplars
and sallows.

Death's Head Hawkmoth
Acherontia atropos

SIZE AND DESCRIPTION Forewing 50–67mm. Upperside with mottled
forewing and pale orange hindwing with two dark bands. Skull-like
markings on thorax make this large moth unmistakable. Larva is up to
130mm long and variably coloured. It is black when young and small,
when it is difficult to detect on the stems of plants, but yellow-green
with dark diagonal stripes when fully grown, the colouration breaking
up the body shape and acting as camouflage in foliage.

HABITAT AND DISTRIBUTION Gardens and fields. Migratory to Europe,
including occasionally to southern England.

FOOD AND HABITS May breed successfully in Europe, but rarely in Britain.
Larvae feed on potato plants and similar.

Bittersweet, a
larval food plant

Caterpillar

Eyed Hawkmoth
Smerinthus ocellatus

SIZE AND DESCRIPTION Forewing 36–44mm. Forewing has a wavy trailing edge; hindwing is pink with a large blue 'eye'. Bright green larva has seven diagonal yellow stripes on each side of its body and a greenish-blue horn at the rear.

HABITAT AND DISTRIBUTION Open woodland, parks and gardens across Europe, but not Scotland or northern Scandinavia.

FOOD AND HABITS Flies May–July, in 1–2 broods. Comes towards light. Larvae feed on willows and apples. Overwinters as a pupa in the soil.

Pine Hawkmoth
Hyloicus pinastris

SIZE AND DESCRIPTION Forewing 35–41mm. Upperside is variable dark grey, and hindwing is darker than forewing. Wing edges are chequered black and white. Larva is at first green with yellow stripes, then darker green with blackish transverse stripes and a wide reddish-brown stripe along the back.

HABITAT AND DISTRIBUTION Pine forests throughout Europe.

FOOD AND HABITS Flies June–July. Larvae feed on Scots Pine and Norway Spruce. Overwinters as a pupa.

Striped Hawkmoth
Hyles lineata

SIZE AND DESCRIPTION Forewing 33–42mm. Upperside has dark olive-brown forewings with white stripes, and pink hindwings bordered with black. Larva is dark green or black with small yellow spots.
HABITAT AND DISTRIBUTION Found in gardens and hedgerows worldwide.
FOOD AND HABITS Flies May–October. One brood a year. Larvae feed on snapdragons and Hedge Bedstraw. Overwinters as a pupa.

Hummingbird Hawkmoth
Macroglossum stellatarum

SIZE AND DESCRIPTION Forewing 20–24mm. Mouse-grey forewings and
a hairy thorax. Hindwings are golden orange. Larva is about 50mm
long, and green with yellow, white and green horizontal stripes.
HABITAT AND DISTRIBUTION Parks, gardens and flowery banks. Southern
Europe, moving northwards in summer and reaching Britain in
varying numbers.
FOOD AND HABITS Day-flying throughout the year. Usually seen in
summer in Britain. Hovers in front of flowers, drinking nectar through
its long proboscis. Larvae feed on bedstraws.

Elephant Hawkmoth
Deilephila elpenor

SIZE AND DESCRIPTION Forewing 28–33mm. Upperside has olive-brown forewings, and pink and brown hindwings. Larva measures up to 80mm in length, and is yellow-brown or green. Its anterior bears a resemblance to a trunk-like snout, hence the name of the moth. Larva has conspicuous large eyespots on segments 2 and 3, which it uses in defence display. When threatened it withdraws its head into its body, causing its front part to swell, rear up and display the eyespots in a snake-like manner.

HABITAT AND DISTRIBUTION Woodland clearings, meadows, gardens, river valleys and waste ground across Europe. Found throughout Britain and Ireland, except for north and east of Scotland.

FOOD AND HABITS Flies June. Larvae feed on willowherbs and bedstraws. Overwinters as a pupa.

**Rosebay Willowherb,
a larval foodplant**

Caterpillar

Privet Hawkmoth
Sphinx ligustri

SIZE AND DESCRIPTION Forewing 41–55mm. Brown wings have black markings, and there is a tan trailing edge to the forewing. Body is striped with pink and black. Larva is up to 100mm long, and green with seven purple-and-white stripes on each side of its body.

HABITAT AND DISTRIBUTION Woodland edges, hedges, parks and gardens across Europe except Ireland, Scotland and far north of Scandinavia.

FOOD AND HABITS Flies June–July, drinking nectar on the wing, especially from honeysuckle. Larvae feed on privets, ashes and lilacs. Overwinters as a pupa in the soil.

Honeysuckle, an adult moth foodplant

Caterpillar

Lime Hawkmoth
Mimas tiliae

SIZE AND DESCRIPTION Forewing 23–39mm. Foreground colour is pale pink to olive-green or brick-red. Central band may be enlarged to form a cross-band, or reduced to a small spot. Larva is large and green, with yellow and red stripes along the body.

HABITAT AND DISTRIBUTION Woods, parks and gardens, and common in towns. Most of Europe excluding Scotland, Ireland and far north.

FOOD AND HABITS Flies May–July. Larvae feed mainly on limes, elms, Downy and Silver Birches, and alders. Overwinters as a pupa.

Buff-tip
Phalera bucephala

SIZE AND DESCRIPTION Forewing to 30mm. Silver-grey wings with an orange tip to each forewing and an orange head, giving it a broken-twig appearance when at rest. Larva is about 45mm long with yellow stripes and sparse hairs.

HABITAT AND DISTRIBUTION Woods, parks, orchards and gardens across Europe except far north.

FOOD AND HABITS Flies May–August. Larvae feed on leaves of oaks, limes, elms and other trees.

Puss Moth
Cerura vinula

SIZE AND DESCRIPTION Forewing 31–40mm. Upperside with white forewing intricately patterned with black. Hindwing is greyish with dark veins. Larva is up to 65mm long, stout and bright green, with a diamond-shaped black saddle, two long tails at the rear and a brown head surrounded by pink. If threatened it raises the front part of its body, displaying the pink ring around the head, raises its tails, from which it extrudes two lash-like red filaments, and may also squirt a jet of formic acid from a throat gland.

HABITAT AND DISTRIBUTION Woods and hedges throughout Europe. Widespread in Britain.

FOOD AND HABITS Flies May–June. Larvae feed on trees such as willows and poplars, particularly suckers or regrowth of Goat Willow and Aspen in sunny positions. Overwinters as a pupa.

Aspen, a larval foodplant

Caterpillar in
defence posture

Figure of Eight
Diloba caeruleocephala

SIZE AND DESCRIPTION Forewing 15mm. Brown-and-grey forewing has a figure-of-eight marking. Hindwing is pale grey-brown. Grey-blue larva is up to 40mm long and has black spots and yellow lines.
HABITAT AND DISTRIBUTION Woodland, scrub and gardens across Europe.
FOOD AND HABITS Flies September–October. Larvae feed on Hawthorn, Blackthorn and other rosaceous shrubs.

Blackthorn, a
larval foodplant

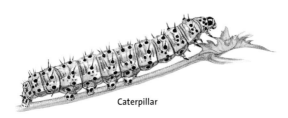

Caterpillar

Great Prominent
Peridea anceps

SIZE AND DESCRIPTION Forewing 26–36mm. Forewing greyish-brown with darker markings of grey, red-brown and ochre. Hindwing is yellowish-white. Female slightly larger than male. There is a melanic form. Larva is yellow-green and has yellow and reddish oblique stripes on its sides.

HABITAT AND DISTRIBUTION Widely distributed and quite common in Europe except for extreme north, northern Scotland and Ireland.

FOOD AND HABITS Flies April–June. One brood a year. Larvae feed on oaks. Overwinters as a pupa.

Gypsy Moth
Lymantria dispar

SIZE AND DESCRIPTION Forewing 24–32mm. Male (shown here) is dark dusky brown, female mainly white. Female does not fly. Larva is pale yellowish with dark grey stripes and tufts of hairs.

HABITAT AND DISTRIBUTION Woodland throughout Europe, but no longer occurs in Britain except as an occasional migrant.

FOOD AND HABITS Flies July–August. Young larvae are dispersed on the wind. Larvae feed on trees, especially oaks and poplars. Overwinters as an egg to hatch the following spring.

Yellow-tail
Euproctis similis

SIZE AND DESCRIPTION Forewing to 20mm. Totally white and rather hairy.
Male has a thinner abdomen than female; both sexes have yellow
tufts at the end. Black larva is up to 40mm long with red stripes and
white spots. It is very hairy, and contact with skin can cause a rash.
HABITAT AND DISTRIBUTION Woods, gardens, parks, orchards, tree-lined
streets and hedges. Most of Europe, but rare in Scotland and Ireland.
FOOD AND HABITS Flies June–August. Larvae feed on Hawthorn,
Blackthorn and fruit trees.

Brown-tail
Euproctis chrysorrhoea

SIZE AND DESCRIPTION Forewing to 20mm. Totally white and rather hairy. Male has a brown abdomen. Female has a white abdomen with brown tufts at the end. Larva is black with a white stripe and yellow tufts of hair, which can cause rashes if touched.

HABITAT AND DISTRIBUTION Woods, hedges, parks and gardens from eastern and southern England across much of mainland Europe.

FOOD AND HABITS Flies July–August. Larvae feed in groups on many species of tree and bush.

Vapourer
Orgyia antiqua

SIZE AND DESCRIPTION Forewing 17mm. Chestnut with a white spot on each forewing. Male (shown here) is winged, but female has only vestigial wings. Larva is up to 35mm long, and dark grey with red spots and four cream tufts of hair on the back. Its body is covered with finer hairs.

HABITAT AND DISTRIBUTION
Woods, parks, gardens,
hedges and tree-lined
streets across Europe.

FOOD AND HABITS Flies
June–October, the male
by day, the female at
night, in 1–3 broods.
Larvae feed on a range
of deciduous trees.
Eggs overwinter.

**Common Lime, a
larval foodplant**

Caterpillar

Common Footman
Eilema lurideola

SIZE AND DESCRIPTION Forewing 15mm. Pale grey forewing is fringed with yellow. Hindwings are pale yellow. Rests with nearly flat wings. Hairy grey larva has black lines on its back and red lines on its sides.
HABITAT AND DISTRIBUTION Hedges, woods and orchards across Europe.
FOOD AND HABITS Flies June–August. Larvae eat lichen.

Jersey Tiger
Euplagia quadripunctaria

SIZE AND DESCRIPTION Forewing 28mm. Upperside has pied chocolate-brown and creamy white forewing, and bright orange hindwing with chocolate-brown blotchy spots. Thorax is brown with yellow sides, and abdomen is bright orange. Larva is dark brown with tufts of paler hairs.

HABITAT AND DISTRIBUTION Open countryside and mountain slopes and valleys. Widespread in central and southern Europe. In England restricted to south-west, although it may be extending its range.

FOOD AND HABITS Flies August. One brood a year. Larvae feed on a variety of plants such as nettles, plantains, Ground-ivy and Bramble. Overwinters as a larva.

Garden Tiger
Arctia caja

SIZE AND DESCRIPTION Forewing 28–37mm. Chocolate-brown forewing has cream patterning. Hindwings are orange with black spots. Very hairy black-and-brown larva is up to 60mm long, and is known as a 'woolly bear'. It is relatively active, especially when searching for a pupation site, and can run very quickly.

HABITAT AND DISTRIBUTION Open habitats, including gardens and scrub throughout Europe.

FOOD AND HABITS Flies June–August. Larvae feed on a variety of plants such as Common Nettle, docks, Hound's-tongue and many garden plants. Overwinters as a small larva.

Broad-leaved Dock,
a larval foodplant

Caterpillar

Buff Ermine
Spilosoma lutea

Size and description Forewing 19mm. Male (shown here) is pale buff to creamy yellow, with a variable broken dark line on the forewing. Female is whiter. Larva is up to 45mm long and has tufts of long brown hairs.

Habitat and distribution Most habitats, but especially common on waste ground and in gardens throughout Europe.

Food and habits Flies May–August. Larvae feed on wild and garden herbaceous plants.

White Ermine
Spilosoma lubricipeda

SIZE AND DESCRIPTION Forewing 17mm. White with more or less sparse black spots. Hairy thorax and black-spotted yellow abdomen. Larva is up to 45mm long, dark brown and very hairy, with a dark red line down its back.

HABITAT AND DISTRIBUTION Hedgerows, gardens, waste ground and other habitats throughout Europe.

FOOD AND HABITS Flies May–August, in 1–2 broods. Adults do not feed, but larvae feed on herbaceous plants including docks and numerous garden plants.

Cinnabar
Tyria jacobaeae

Size and description Forewing 18mm. Upperside with a very dark greyish-black forewing patterned with scarlet streaks and spots; hindwing is scarlet. Larva is up to 30mm long, slender, and banded with black and yellow. Bright colours warn potential bird predators that it is distasteful. Larvae also live together and feed in large groups, and this makes them easy to recognize and therefore unlikely to be eaten accidentally.

Habitat and distribution Waste ground, newly disturbed ground, roadsides, meadows and heathland. Widespread in Europe.

Food and habits Flies May–July. One brood a year. Night-flying but easily disturbed from daytime resting place in low herbage. Larvae feed on ragworts and groundsels. Overwinters as a pupa.

Common Ragwort, a larval foodplant

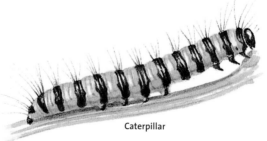

Caterpillar

Pine Processionary
Thaumetopoea pitycampa

SIZE AND DESCRIPTION Forewing 23mm. Upperside with a greyish-white forewing that has a dark brown leading edge, broken dark margin and dark bands. Hindwing is pale with a single dark spot on the rear edge. Larva is grey-black with white hairs and red-brown warts; hairs are extremely irritating to skin if touched.

HABITAT AND DISTRIBUTION Forests in Mediterranean Europe. Rare immigrant to Britain.

FOOD AND HABITS Larvae feed on pine needles. They live communally in large silken tents nestled among pine twigs. Caterpillars move in single file in a long line to a new feeding sites. A serious forest pest.

Garden Dart
Euxoa nigricans

SIZE AND DESCRIPTION Forewing 18mm. Upperwing with a grey-brown forewing and pale whitish hindwing. Larva is ochreous brown and has short hairs.

HABITAT AND DISTRIBUTION Farmland, gardens, marshy areas, commons and waste ground throughout Europe.

FOOD AND HABITS Flies July–August. One brood a year. Larvae feed on a wide variety of plants such as plantains, clovers and docks, as well as lettuce. Overwinters as an egg.

Heart and Dart
Agrotis exclamationis

Size and description Forewing 20mm. Background colour varies from greyish-brown to deep brown. Wings have vaguely heart-shaped and dart-shaped markings. Larva measures about 40mm in length and is a dull brown-and-grey 'cutworm'.

Habitat and distribution Almost any habitat, especially if it is cultivated. Throughout Europe.

Food and habits Flies May–September at night, in 1–2 broods. Larvae feed on stems of herbaceous plants during the night and hide in soil during daytime.

Large Yellow Underwing
Noctua pronuba

SIZE AND DESCRIPTION Forewing 25mm. Varies from pale to dark brown. Hindwings are deep yellow with a black border. The yellow flashes when the moth takes flight, which is thought to confuse predators. Green larva measures up to 50mm in length and has two rows of dark markings on its back.

HABITAT AND DISTRIBUTION Well-vegetated habitats throughout Europe except far north.

FOOD AND HABITS Flies June–October. Flight is fast and erratic. The yellow flashes shown in flight become invisible the moment it lands.

Chinese Character
Cilix glaucata

SIZE AND DESCRIPTION Forewing 15mm. Upperside is whitish; forewing is bordered grey, with spots near the outer edge and a broad brown mark in the centre. Hindwing has a grey outer border. Larva is reddish-brown with black lines.

HABITAT AND DISTRIBUTION Woodland edges and hedgerows. Widespread in central and southern Europe.

FOOD AND HABITS Flies May–June and July–August. Two broods a year. Larvae feed on Blackthorn, Hawthorn, and fruit trees such as plums and pears. Overwinters as a pupa.

Setaceous Hebrew Character
Xestia c-nigrum

SIZE AND DESCRIPTION Forewing to 20mm. Greyish-brown to chestnut with a purplish tinge. There is a pale patch on the leading edge of the forewing. Hindwing is pale and more uniform in colour. Larva is initially green, before becoming pale greenish-grey.

HABITAT AND DISTRIBUTION Lowland areas including cultivated regions, woodland and marshes throughout Europe except far north.

FOOD AND HABITS Flies May–October. Two or more broods a year. Larvae feed on wide variety of herbaceous plants. Overwinters as a larva or pupa.

Mouse Moth
Amphipyra tragopoginis

Size and description Forewing 15mm. Dark brown with three dark spots on each forewing. Underwings are pale. Holds wings flat along its abdomen when at rest. Larva is green with narrow white lines.

Habitat and distribution Widespread in woods, hedgerows, gardens and open countryside with scrub throughout Europe.

Food and habits Flies June–September. Roosts by day in outbuildings, under loose bark and in hollow trees. If disturbed scuttles off in a mouse-like fashion. Larvae feed on plants such as Salad Burnet, Fennel and Hawthorn.

Cabbage Moth
Mamestra brassicae

SIZE AND DESCRIPTION Forewing 18mm. Mottled greyish-brown with rusty scales. Plump larva, up to 50mm long, is brownish-green with subtle dark and pale markings.

HABITAT AND DISTRIBUTION Almost any habitat, but most common on cultivated land throughout Europe except far north.

FOOD AND HABITS Flies throughout the year, but mainly May–September. Larvae feed on cabbages and other herbaceous plants. Overwinters as a pupa.

Clay
Mythimna ferrago

SIZE AND DESCRIPTION Forewing 15mm. Colour varies from straw to reddish-brown, with a white mark in the middle of each forewing. Wings lie flat when at rest. Pale brown larva is marked with thin yellow lines.

HABITAT AND DISTRIBUTION Common in grassy places across Europe.

FOOD AND HABITS Flies May–August. Larvae feed on grasses and other low-growing plants.

Green Arches
Anaplectoides prasina

SIZE AND DESCRIPTION Forewing 20mm. Greenish forewings have variable black markings, while hindwings are dark grey or brown. Larva is brown with darker markings.

HABITAT AND DISTRIBUTION Deciduous woodland over most of Europe.

FOOD AND HABITS Flies mid-June–mid-July at night. Larvae feed on a range of plants, especially honeysuckle and Bilberry.

Common Wainscot
Mythimna pallens

SIZE AND DESCRIPTION Forewing 15mm. Pale-coloured with a creamy forewing and even paler hindwing. Larva is ochreous with three white lines.

HABITAT AND DISTRIBUTION Grassland including meadows and marshes. Occurs widely throughout Europe, including Britain.

Food and habits Flies June–October. Two broods a year. Larvae feed on grasses such as Cock's-foot and Couch Grass. Overwinters as a larva.

Old Lady
Mormo maura

SIZE AND DESCRIPTION Forewing to 35mm. Patterned dark brown and black, said to resemble an old lady's shawl. Greyish-brown larva, up to 75mm long, has dark smudgy diamonds and a broken white line running down its back.

HABITAT AND DISTRIBUTION Woods, hedges, gardens, parks and damp places in southern, central and western Europe, including Britain, although rare in Ireland and Scotland.

FOOD AND HABITS Flies July–August, often coming towards light. Larvae feed on a variety of trees and shrubs. Hibernates as a small larva.

Burnished Brass
Diachrisia chrysitis

SIZE AND DESCRIPTION Forewing to 20mm. The two metallic marks on the forewings vary from emerald to deep gold. Wings are held above the abdomen when at rest. There is a prominent tuft on the thorax. Larva is up to 35mm long and bluish-green, with diagonal white streaks across the back and a white line along the sides.

HABITAT AND DISTRIBUTION Gardens, parks, hedges and waste ground throughout Europe.

FOOD AND HABITS Flies May–October, in 1–3 broods. Larvae feed on nettles and mint. Hibernates over winter in larval form.

Mullein
Cucullia verbasci

SIZE AND DESCRIPTION Forewing 20–25mm. Colour varies from pale straw to mid-brown and is darker streaked. With the wings held close to its body, the resting moth resembles a twig. Larva is up to 60mm long and creamy white, with yellow and black spots.

HABITAT AND DISTRIBUTION Woodland edges, scrub, river banks, gardens and parks across most of Europe, but not Scotland, Ireland and northern Scandinavia.

FOOD AND HABITS Flies April–June. In June and July, larvae feed on mulleins, figworts and buddleias. Overwinters as a pupa.

Grey Chi
Antitype chi

Size and description Forewing 15mm. Mottled grey wings with a small but distinct dark cross in the middle of the forewing, which may also be anvil- or bar-shaped, or reduced. Bluish-green larva has green-edged white lines along its body.

Habitat and distribution Gardens, grassy places and moorlands in much of Europe except far north.

Food and habits Flies August–September, resting on walls and rocks during the day. Larvae feed from April to early June on low plants such as docks and sorrels. Overwinters as an egg.

Merville du Jour
Dichonia aprilina

SIZE AND DESCRIPTION Forewing 23mm. Upperside has forewing finely patterned with white, yellow, pale green and dark brown. Hindwing is greyish with a dark border. Larva varies from red-brown to green-brown to grey-brown, with a broken white line down the centre.
HABITAT AND DISTRIBUTION Parkland and oak woodland. Widespread in Europe, but local. Widespread in Britain and most common in southern England.
FOOD AND HABITS Flies September–October. One brood a year. Larvae feed on oaks, first on buds, then on leaves. Overwinters as an egg.

Angle Shades
Phlogophora meticulosa

SIZE AND DESCRIPTION Forewing 25mm. Varies from brown to green, with distinctive V-shaped markings. Forewing's trailing edge has a ragged look, which is exaggerated by the moth's habit of resting with its wings curled over. Fat green larva is up to 45mm long and has a white line (often faint) along its back.

HABITAT AND DISTRIBUTION A migrant that can be found found in almost any habitat in Europe.

FOOD AND HABITS Flies most of year, but mainly in May–October. Larvae feed on a variety of wild and cultivated plants. Overwinters as a larva.

Silver Y
Autographa gamma

Size and description Forewing 20mm. Varies from purple-tinged grey to almost black, with a silver Y-mark on the forewing. Green larva is up to 25mm long.

Habitat and distribution A migrant found all over Europe. Breeds all year in southern Europe. British and other northern breeders do not survive winter and are supplemented by migrants.

Food and habits Flies by day throughout the year. Attracted by nectar. May be seen in autumn alongside butterflies on buddleias. Larvae feed on low-growing wild and cultivated plants.

Grey Dagger
Acronicta psi

SIZE AND DESCRIPTION Forewing to 20mm. Pale to dark grey, with dark and apparently dagger-shaped marks on the forewings. Hairy grey-black larva has a yellow line along its back, red spots along its sides and a black horn on its first abdominal segment.

HABITAT AND DISTRIBUTION Woodland, commons, parks and gardens across Europe except far north.

FOOD AND HABITS Flies May–September, with larva feeding August–October on wide range of broadleaved trees. Overwinters as a pupa.

Red Underwing
Catocala nupta

SIZE AND DESCRIPTION Forewing to 35mm. Grey-mottled forewings make the moth well-camouflaged on tree-bark, but bright red underwings are very conspicuous in flight. Pale brown larva has warty bud-like lumps on its back.

HABITAT AND DISTRIBUTION Woodland, hedges, gardens and parks across Europe except northern Scandinavia.

FOOD AND HABITS Flies August–September at night. Flies erratically, flashing its red underwings to confuse predators. Larvae feed in May–July on willows, poplars and aspens.

Herald
Scoliopteryx libatrix

SIZE AND DESCRIPTION Forewing to 25mm. Purplish to orange-brown with bright orange scales near the head. Trailing edge of the forewing is ragged. Slender green larva is up to 55mm long and has thin pale yellow lines along its sides.

HABITAT AND DISTRIBUTION Woodlands, gardens, parks and open countryside across Europe except northern Scandinavia.

FOOD AND HABITS Flies August–October, and in spring after migration. Larvae feed on willows and sallows. Overwinters as an adult.

Snout
Hypena proboscidalis

SIZE AND DESCRIPTION Forewing 19mm. Upperside has a dark grey-brown forewing with a paler hindwing. The long palpi (sensory organs on the head) give this species its common name. Larva is long and slender, yellow-green to dark green, with white lines along the back and sides.
HABITAT AND DISTRIBUTION Gardens, waste ground, open woodland and hedgerows throughout Europe.
FOOD AND HABITS Flies June–early August and September–early October. Larvae feed on nettles. Overwinters as a larva.

Index

Common names

Scientific names